Advanced Treasure Hunting with the Fisher QuickSilver Series Metal Detectors

By

Andy Sabisch

FISHER m-SCOPE ®

FISHER RESEARCH LABORATORY

Dedication

This book is dedicated to my wife, Roseanne, and two children, Paul and Leigh. Their love and support for my interest in treasure hunting have kept me involved in all aspects of this exciting sport over the years and helped keep me motivated during the writing of this book. Without them, most of the material contained in the book never would have been written.

First Printing, August 1995
Second Printing, June 1996
Third Printing, February 1999

Library of Congress Catalog Card Number: 95-060828

ISBN: 1-883170-05-2

FRL870539B

Published by:

FISHER RESEARCH LABORATORY
200 W. Willmott Rd., Dept. ATH
Los Banos, CA 93635
TEL: (209) 826-3292; FAX: (209) 826-0416
www.fisherlab.com; email: info@fisherlab.com

FOREWORD

Several years ago, we published a 44-page book titled "Advanced Nugget Hunting with the Fisher Gold Bug Metal Detector." We printed what I thought would be a 5-year supply for a pretty small target audience. But we sold out within a year and the book is now in its third printing.

We followed up with two more books: one on the 1280-X Aquanaut and one on the Gemini-3 "two-box" metal detector. Once again we underestimated our potential readership, and both books are destined for multiple reprints.

All three books were written for the owners of specific metal detectors, but we've learned that treasure hunters are hungry for good information no matter what kind of metal detector they own.

This book, by Andy Sabisch, is the fourth in the series on Fisher detectors and perhaps my favorite since I'm hooked on using the CZ QuickSilver models. After years of testing them, giving seminars about them and using them for just plain fun, I'm still learning new tricks. If you're a fellow CZ user, or if you just want some good, solid tips on treasure hunting, read on . . .

Good hunting,

Jim Lewellen
President and General Manager
Fisher Research Laboratory

Contents

** Chapter 9 was written by David Toussaint, Fisher marketing publications writer.*

INTRODUCTION

Having been involved in treasure hunting for nearly 30 years, I have seen the industry's technology reach levels that would have been hard to imagine even a few short years ago. Revolutionary features that most treasure hunters now take for granted—such as

The author, Andy Sabisch, hunts for old coins and other lost teasures near an old house with his Fisher M-SCOPE CZ-20 metal detector.

discrimination, target identification, depth reading and ground balance—were not even on the drawing boards back in the late 1960s. Looking back, it's hard to imagine that any of us actually found anything with the crude equipment we had at the time.

Fisher Research Laboratory, the industry's oldest metal detector manufacturer, has been building detectors since 1931 for both commercial and recreational users. Many of the features found on virtually all metal detectors today—such as silent-search, slow-motion VLF discrimination, the S-shape shaft design, and "turn-on-and-go" circuitry —were originally developed by Fisher engineers. Employing some of the field's most-talented electrical and electronic engineers, Fisher's goal has always been to develop the finest metal detectors using state-of-the-art technology.

While this book focuses on Fisher's new QuickSilver line of metal detectors, it also provides experienced as well as novice treasure hunters tips and techniques that can improve their success in the field, regardless of the particular brand of metal detector they decide to use. This book covers a wide spectrum of various forms of treasure hunting: coin hunting, relic hunting, beach and shallow-water hunting, diving, electronic prospecting, and even competition hunting. Readers should be able to find several chapters that pertain to their specific interests.

In closing, I would like to offer my thanks and appreciation to the following treasure hunters who took the time to contribute their experience and expertise to this book. Many of the tips and techniques provided throughout this book would be missing without their input.

Jerry Denton	Larry Dunnagan	Roy Gouge
Larry Hayes	Kevin Reilly	John Sanders
David Sponenberg Sr.	Dave Sponenberg	
George Streeter	Amos Taylor	

Good luck and good hunting with your QuickSilver!

Andy Sabisch

CHAPTER 1
THE QUICKSILVER'S CONTROLS

Fisher engineers spent several years and literally hundreds of hours field testing various circuit designs before the CZ-6 QuickSilver was first introduced. Quickly gaining a reputation as

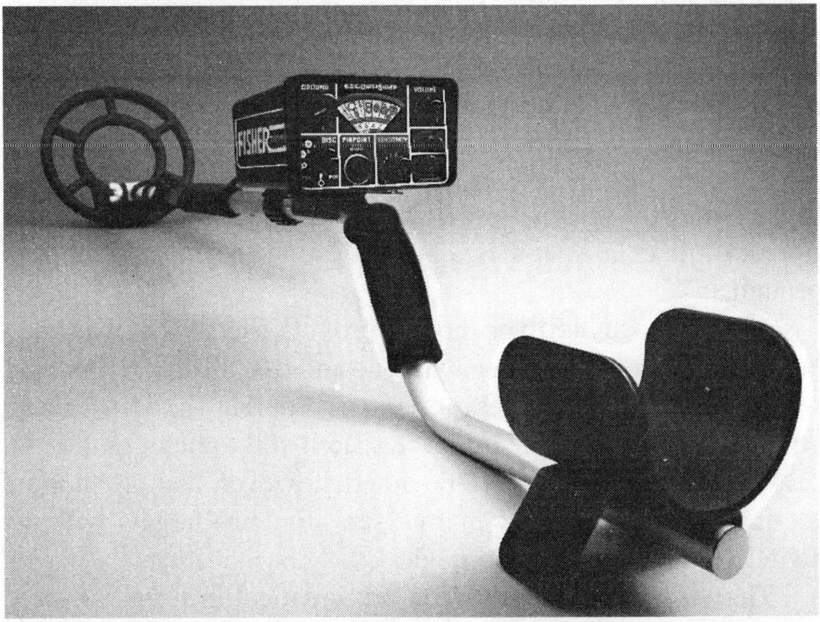

The Fisher CZ-6a QuickSilver is widely used in all facets of treasure hunting, including coin hunting, relic hunting and beach hunting.

a detector that could locate and accurately identify deeply buried targets, the CZ-6 became widely used in all facets of treasure hunting, including coin hunting, relic hunting and beach hunting. The CZ-5, which incorporates virtually all of the performance features found in the CZ-6, followed, and then the submersible version of the QuickSilver, the CZ-20, was released.

Based on discussions with detectorists and metal detector dealers from around the world, the Fisher design team developed two distinct goals during the design and development phase of the

CZ-6. First, its performance in terms of detection depth and discrimination had to be top-notch. Secondly, and equally as important, was the need for its operation to be simple and straightforward. Feedback from most of the treasure hunters they spoke to indicated that the majority of the detectors on the market were too complicated for most users to realize the true capabilities of the unit.

A design feature common to all of the QuickSilver detectors, which contributes to their exceptional performance in virtually any type of ground, is the Fourier Domain Signal Analysis System. Transmitting two ground-balanced signals at different frequencies (5 kHz and 15 kHz) simultaneously, more-accurate target identification and improved detection depth is achieved, even in highly mineralized ground.

NOTE: The CZ-6 and 6a are splash proof but have not been designed to be totally submerged for such uses as shallow-water hunting or treasure diving.

All of the QuickSilver models provide users with two independent search modes as well as a no-motion, all-metal pinpoint mode. Both search modes utilize the Fourier Domain Signal Analysis circuitry and require the search coil to be in motion to respond to targets. It should be noted, however, that the motion required is so slight that many users pinpoint targets without switching to the no-motion mode.

The CZ-5, CZ-6 and CZ-6a are identical in terms of their performance. The difference is in the control housing construction and connectors used for the search coil cable and headphone jack. The CZ-6 and 6a have been designed to be water resistant; i.e., they can be used in the rain, ocean surf or anywhere where water could splash on the control housing and damage a conventional land detector. The CZ-6 and 6a have not been designed to be totally submerged when shallow-water hunting or treasure diving.

The CZ-6a replaced the CZ-6. The main difference between the two models is the type of connector used on the search-coil cable. All of the various search coils available for the CZ-5 are now interchangeable with the CZ-6a. (The CZ-6 coils are not

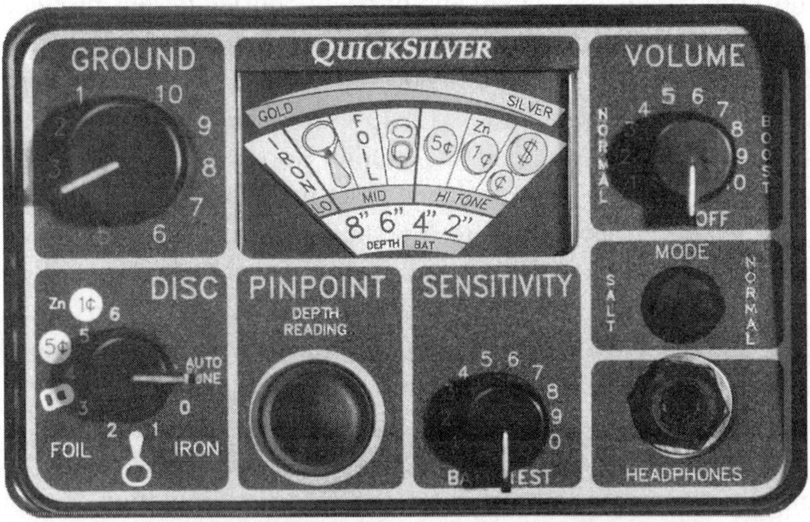

The Fisher CZ-6a QuickSilver control panel features an object-oriented target-I.D. gauge that measures target depth and groups targets as iron, pulltab, foil, nickel, zinc penny and silver.

interchangeable with the CZ-5 and 6a).

Another difference between the CZ-5 and CZ-6/6a is that due to the water-resistant design of the CZ-5/6a control housing, there is no internal speaker. Fisher's engineers opted to eliminate the speaker on the CZ-6 and 6a since most treasure hunters reported that they used headphones almost exclusively. However, an optional plug-in speaker is available (see chapter 8).

The manuals provided by Fisher are well written; however, a brief overview of the controls is warranted at this time to make sure that users fully understand the tips and techniques given throughout the rest of the book.

The CZ-5, 6 and 6a are controlled by four knobs, a push-button and a toggle switch, all located on the face of the control housing. The CZ-20 has the same controls as the CZ-5, 6, and 6a with the exception of the MODE toggle switch. The photograph above shows the CZ-6a control panel, which except for its water-resistant construction is identical to the CZ-5.

The volume control on all four models serves a dual function: it turns the unit on and adjusts the volume of the audio signal heard through the speaker or headphones. The QuickSilver has a built-in signal booster, which is activated via the volume control. On most other detectors, small and/or deeply buried targets produce weak signals that are frequently missed by treasure hunters. Unfortunately, they are usually the older and more valuable items we are all hoping to find. To help users detect these marginal signals, Fisher developed the Audio Boost circuit. Once the volume control is turned past the "5" setting, the volume of signals from shallow (less than 4 to 5 inches) or large targets remains constant while the audio response from faint signals is amplified to ensure easy detection. The first time you search a "hunted-out" park and receive a clear, discernible signal from a Barber dime at 9 or 10 inches, you will see just how effective this feature can be.

The discrimination control also incorporates a design unique to the QuickSilver series. Unlike other detectors that utilize an infinitely variable potentiometer, the CZ's discrimination control has only seven distinct settings. Labeled "0" through "6," each setting is designed to reject the specific type of target identified on the control panel face.

Another difference between the CZ's discrimination circuitry and that found on most other brands is that the level of discrimination selected has little, if any, affect on the CZ's detection depth. Other detectors often lose a considerable amount of actual detection depth as increased levels of discrimination are used.

All of the QuickSilver models feature an audio target identification system that enables the user to quickly determine if targets are worth recovering. There are three distinct audio signals that are produced depending on the target's composition. Iron items produce a low tone (200 Hz); pull tabs, tin foil, screw caps, and most gold rings a medium tone (450 Hz); and coins as well as silver jewelry a high tone (1 kHz).

In addition to the audio target identification circuitry, the CZ-5, 6 and 6a feature a meter that not only provides a visual representation of the probable target detected but also gives the target depth when the pinpoint button is depressed. Realizing that a target too large or close to the search coil may overload the discrimination circuitry and prevent the detector from accurately identifying the target, Fisher's engineers developed an "Overload Indication" function. Simply stated, when an overload signal is received, the CZ responds with a distinctive bell-tone audio signal, alerting the operator to the fact that it could not identify the target. If this happens, simply raise the coil an inch or two off the ground and recheck the target. A small, shallow target will produce a solid-tone I.D. and a lock-on meter indication. A large, shallow target will continue to produce a bell tone several inches off the ground.

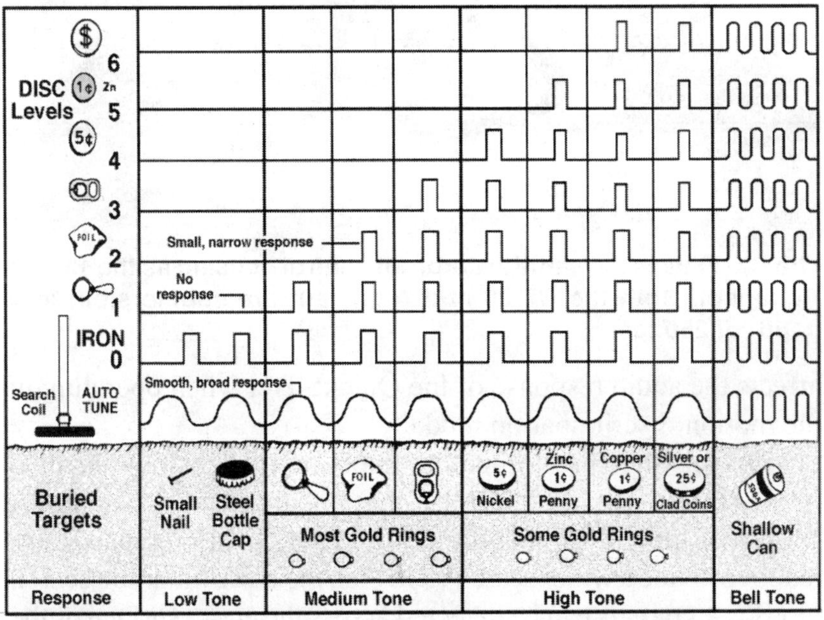

Response of various targets based on individual discrimination settings.

> **NOTE: While the CZ-20 lacks the meter found on the CZ-5, 6 and 6a, the audio response is the same for any given discrimination setting.**

While users may initially question the usefulness of the CZ's meter because of the limited number of segments (7) that it contains, they will quickly find that its accuracy is above average, and it prevents the user from missing a valuable target due to its orientation or proximity to a trash target. On most other target-I.D. detectors, a coin on edge or one buried next to a piece of trash will not produce a meter reading typical of the coin by itself, and as a result, most treasure hunters pass these signals up and leave many good targets for the next hunter to find.

The chart above shows how the discrimination-control setting

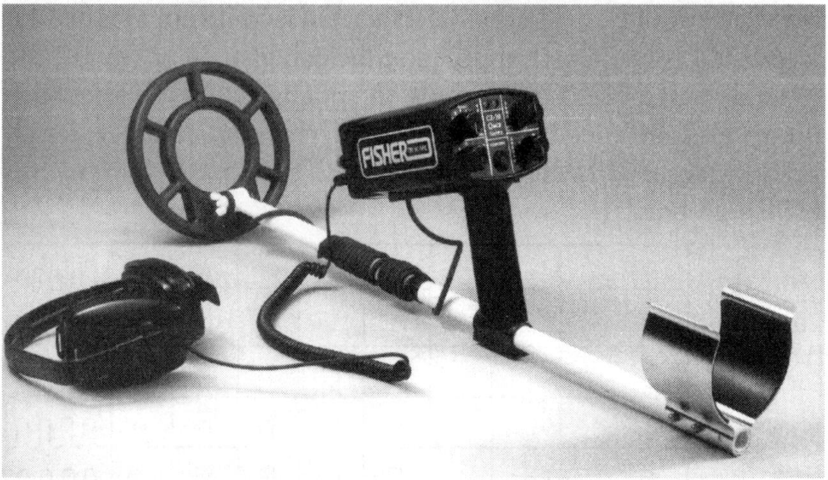

The CZ-20 uses the same waterproof control housing as the 1280-X Aquanaut. Both the CZ-20 and Aquanaut are submersible to a depth of 250 feet.

affects the audio response of the QuickSilver when operating in the motion-discrimination mode.

Placing the discrimination knob in the position labeled AUTOTUNE activates the VLF-motion, all-metal search mode. While this mode does not provide any discrimination capabilities, it does provide a wider search field than the discrimination search mode (i.e., targets can be detected across the entire surface of the coil) as well as a slight increase in overall detection depth and sensitivity to smaller targets. As will be discussed in subsequent chapters, many treasure hunters prefer to hunt in this mode for the improved performance it affords and then simply click the discrimination control to the "0" position to accurately identify targets prior to recovery.

Here's a word of caution regarding the QuickSilver's discrimination circuit (or for that matter ANY detector's discrimination circuit): use it sparingly. Having been involved in treasure hunting for nearly 30 years, I can remember the days when discrimination was only a designer's dream, and trash filled one's pouch at the end of the day—often outnumbering good targets by a 4- or 5-to-1 margin. While the addition of discrimination circuits has greatly reduced the amount of worthless targets metal detectorists recover, overuse of this control is probably the number one reason why so many good targets still remain in the ground. Unfortunately, due to the wide variation in target size, composition and orientation, metal detectors often "see" common trash targets, such as pull tabs, tin foil and screw caps as being similar to good items, making rejection of all trash quite difficult.

Use only enough discrimination to eliminate those trash targets that are being found in sufficient quantity to make recovering them an ineffective use of your time. If you are searching a picnic area and are finding a large number of rectangular pull tabs but little other trash, leave the discrimination control at "1" and use the meter to visually reject the bothersome target rather than increase the discrimination control to "4." This will prevent you from missing a potentially valuable target, such as a gold ring that might register in the FOIL or RING TAB regions of the meter, since these targets will be rejected at discrimination settings below "4."

The sensitivity control provides three separate functions. First, when searching in the motion-discrimination mode, it is used to vary the strength of the signal being transmitted from the coil. Second, in the AUTOTUNE mode, the sensitivity control is used to set the audio threshold heard through the speaker or headphones. And third, when turned fully counterclockwise, it activates the built-in battery-check circuitry. Battery strength is indicated by the volume of the audio signal produced as well as the meter indication on the CZ-5/6/6a and the intensity of the two LEDs on the CZ-20.

Besides the discrimination control, the sensitivity control is probably the most-abused control on a metal detector. The QuickSilver has been designed with more than enough sensitivity for most areas, and as a result, turning the control up too high will actually result in a *loss* of detection depth and erratic operation caused by signal overload.

Think of the sensitivity control on your detector as the dimmer switch on your car's headlights. On a clear night, the high-beam setting will allow you to see much farther than the low beam setting; however, on a foggy night using the high beams will actually reduce your visibility because light is reflected back from the water particles in the air. Mineralized ground will have the same effect on the signal from your detector if the sensitivity (power output) is set too high. For maximum penetration and target recognition, the sensitivity control should only be set as high as conditions will allow.

To determine the optimum setting for this control, place the sensitivity control near the maximum position ("10") and sweep the coil across the ground to be searched. If you receive more than just an occasional chirp or false signal, reduce the setting a small amount and repeat the test. The ideal setting is one at which you may receive a false signal every few sweeps. Don't worry if you find the control set near the low end of its adjustment range—if that is as high as the ground conditions will allow, you will still find more than someone who runs their sensitivity too high and misses targets due to signal overload. I have personally recovered coins and small Civil War relics at depths of up to 8 inches with my CZ even with the sensitivity set at its minimum position.

The final control knob common to all of the QuickSilver models is labeled GROUND and is possibly the most important control on the detector. Mineralization, which is present in all soil to varying degrees, greatly affects a metal detector's detection capabilities. In highly mineralized soil, some detectors may be rendered virtually useless in locating targets buried even a few inches deep.

While most treasure hunters realize that proper adjustment of the ground-balance circuitry is crucial to achieve maximum detection depth, this is probably the least understood and under-utilized adjustment on a detector today. For this reason, many manufacturers have eliminated the ground-balance adjustment. Their detectors come with a preset ground-balance setting designed to work under most conditions. The problem users face with this design is that when searching areas with higher-than-normal levels of mineralization, detection depth will be adversely affected when compared to a manual ground-balance detector under the same conditions. Recognizing that a simple-yet-effective manual ground-balance circuit was necessary if the CZs were going to work effectively in all parts of the world, Fisher engineers developed what has become one of the easiest-to-adjust ground-balance circuits in the industry.

There are two methods that can be used to ground balance the CZ. They are both extremely simple and easy to master. The QuickSilver instruction manuals devote an entire section to the two techniques dubbed the "push-button" and "bobbing" methods. Even a complete novice will be able to accurately adjust the CZ for even the most-severe mineralization after a short practice

The battery compartment of the CZ-20 is totally separate from the electronics compartment, which eliminates any possibility of damaging the electronics if the batteries leak or the battery compartment floods. The tightness of the battery door and condition of the O-rings should be checked each time you take the unit into the water.

session.

The toggle switch labeled SALT/NORMAL is only found on the CZ-5, 6 and 6a. On other VLF-type detectors, approaching the wet sand/surf region of an ocean beach usually results in erratic operation and false signals caused by the conductivity of wet salt. As a result, treasure hunters either stay away from these areas or are forced to decrease their detector's sensitivity to the point where they lose almost all of their detection capabilities. The SALT/

NORMAL circuit on the QuickSilvers has eliminated this problem. When you approach the wet-sand area, simply switch to the SALT setting, reground balance the detector, and continue hunting. On most beaches, very few false signals will be experienced even with the sensitivity control set at the mid position ("5"). While the CZ-20 lacks the SALT/NORMAL switch, the SALT mode on the CZ-20 is active at all times since many CZ-20 users

The CZ-20 identifies targets with its discrimination control. At a setting of "1," it tunes out iron, at "2" it tunes out round pull tabs, at "3" it tunes out rectangular pull tabs, at "4" nickels, at "5" zinc pennies and at "6" silver coins.

will be hunting mainly in salt water or on saltwater beaches. Although the SALT mode seems to work as well as the NORMAL mode in most soils (better in some), some users may notice that the NORMAL mode is more sensitive to some small gold targets.

One feature of the CZ-20 that is not really described in the

This decal, attached to the bottom of the CZ-20 control housing, serves as a handy reference chart for CZ-20 target-I.D. discrimination settings.

owner's manual is the dual purpose of the two LEDs located in the center of the control panel. They indicate battery strength when the battery-check circuit is activated (by their intensity) and also provide the user with an indication of relative signal strength when using the pinpoint mode. A weak signal will cause only the left LED to light. As the signal strength increases, its brightness will also increase until the second LED comes on. The LEDs help divers and water hunters accurately pinpoint targets, especially when used in conjunction with the VCO audio response heard through the waterproof headphones.

Another design feature of the CZ-20 that deserves mention is the control housing. Similar to that used on the Fisher 1280-X Aquanaut or Impulse, it has developed an enviable reputation of being both rugged and leak-resistant.

The battery compartment is physically separated from the electronics section. This ensures that in the event the battery-compartment door leaks, no water can reach the circuit board and damage the detector. As a testimonial to how this design can reduce downtime, several treasure divers I know have reported that they have flooded the battery compartments of their CZ-20s by neglecting to check the tightness of the battery door before entering the water. After simply rinsing out the compartment, checking the condition of the special battery holder and replacing the batteries, the detector worked perfectly.

One important note, however, is that the battery holder used on the CZ-20 has been designed specifically for that unit and must be purchased through a Fisher dealer. If you plan to spend much time in the water with your CZ-20, you should probably purchase an extra battery holder (Fisher part number 2037582) to minimize any downtime caused by a flooded battery compartment. This feature, unique to Fisher's line of underwater metal detectors, shows that their engineers understand and have addressed the adverse factors that face active treasure hunters.

CHAPTER 2
FAMILIARIZING YOURSELF WITH YOUR DETECTOR'S PERFORMANCE

Bench-testing your CZ will let you see how varying the control settings affects its overall performance.

Most treasure hunters, even those who have been hunting for years, tend to set their detector's controls at one point and then never change them. The settings may have allowed them to be successful in a certain application (such as hunting the local park, school yard, or Civil War campsite) or at a specific location. As a result, they rarely check to see if a different setting would be better in another situation.

The purpose of this chapter is to provide you with several techniques that will give you a better understanding of how each of your CZ's controls affect its operation and how to maximize its performance under all types of conditions.

Bench Testing Techniques

Years ago, virtually every metal-detector manufacturer included tables in their literature showing the detection depth of various targets in air. While these values were often quite impressive, they rarely indicated how well the detector actually performed in the field. Ground mineralization, electrical interference and the skill of the operator all tended to reduce the actual detection depth achieved when searching for buried objects. Because of the negative feedback received from treasure hunters and even dealers on this somewhat deceptive practice, air-test information is rarely used to promote the detection capabilities of today's metal detectors.

Even though air tests do not provide an accurate indication of how a particular detector performs in the field, they do allow the user to see how the detector will respond to various targets at different discrimination and sensitivity settings. To conduct your own air tests, the first thing you should do is to assemble the following items:

- Various samples of the type of items you will be searching for; i.e., coins, jewelry, artifacts, etc.
- Various samples of trash targets you can expect to find in your searches; i.e., pull tabs, tin foil and screw caps for coin hunters—iron nails for relic hunters.
- A wooden ruler to measure detection-depth capabilities for each type of target
- Several copies of the table shown on page 17 to record your results.

Next, place the QuickSilver on a table with the search coil facing away from any metal, such as a lamp, bureau or chair. Remember that the coil will detect metal both above it and below it, so make sure the table itself does not contain any metal that will

Air Test Response Form

Test Item	Disc. Setting	Sens. Setting	Audio/Meter Response (Y/N)	Max. Detection Depth

interfere with the test.

For each item to be tested, start by placing the discrimination control in the AUTOTUNE position and setting the sensitivity control at "1." As you wave each target in front of the search coil, record the maximum depth at which you receive a consistent, clear signal. For the purpose of these tests, do not continue to move the target beyond the point at which a consistent signal can not be obtained. Test each item at sensitivity settings of "1," "4," "7" and "10" before changing the discrimination setting. Repeat the test at these settings for each discrimination position.

When you have completed air test for each of the items you collected, you will have a wealth of useful information that will help you improve your overall success rate in the field. While the owner's manual provides a guide to where specific targets will be accepted or rejected based on the discrimination-control setting, your own data will also show you how the sensitivity control affects target acceptance and detection depth. In addition, the data will be unique to your detector. Keep in mind that minor tolerances in the process used to manufacture the circuit boards in metal detectors may affect how an individual detector responds to various targets, and by performing your own air test as described above, you will know exactly how your detector will respond in the field.

Once you have bench-tested your detector using targets similar to those for which you will be searching, you are ready to take your detector outside and see how well it performs in the field.

The Test Garden

Do you really know how your detector will respond to a silver dime 6 inches deep, a coin on edge or a thin gold ring in mineralized ground? Surprisingly, many treasure hunters, even those who have been hunting for years, can only surmise the answers.

In talking with a number of successful treasure hunters from all parts of the country, one of the most important factors to which they have all attributed their success in the field is their familiarity with what their detector tells them when it responds to a buried target. Being able to distinguish the faint whisper of a thin gold ring at the edge of your detector's range from the chirp caused by ground mineralization or trash targets is the most important skill a treasure hunter can acquire when it comes to increasing success in the field. Most treasure hunters are not aware of how changes in ground conditions can affect a detector's response to buried targets. As the ground dries out, the signal from the coil does not penetrate as effectively, which often results in a weak or broken signal that may have been quite clear and pronounced had the ground been damp. Learning how to detect good targets located adjacent to trash is another skill that, once learned, will enable you to find "keepers" in areas that others have given up on long ago.

Most of the coins and other valuable targets being recovered today are lying flat in the ground, providing a large surface area to which the coil can react. If the target is buried at an angle or even on edge, it becomes much more difficult for the detector to see, and as a result, most treasure hunters will miss it and continue on with their search. Again, if you learn what type of signals these targets produce, the empty holes in your collection will quickly be filled.

A test garden is extremely simple to construct and can be as basic or elaborate as you want. The first step in putting together a test garden is to select a representative sample of the type of targets you will normally be searching for in the field. If you split your time between two or more different types of treasure hunting; such as coin hunting, beach hunting or relic hunting; you might want to consider building a separate test garden for each type of hunting.

The accompanying sketch shows one of the test gardens I constructed for my wife and myself to practice our coin-hunting techniques. You should select items to be included in your test garden based on those targets that frequently give you problems in the field. Deeply buried coins, various types of pull tabs, jewelry items and different trash targets are good choices because of the difficulty in either rejecting or detecting them under actual field conditions.

Once you have collected the items for the test garden, select an area of your yard to bury them in. Before actually burying any of the items, take your detector, carefully search the area in the AUTOTUNE mode and remove any metal objects present. Draw a sketch of the planned test garden, indicating what each item is and the depth at which it will be buried. Use a ruler to make sure they are all buried at the proper depth, and then mark the exact center of each target with a colored golf tee. On the top of each golf tee, paint a number that corresponds to the number of each target on the drawing of the test garden target.

You will find that the amount of sensitivity or discrimination used plays a big part in how well the detector responds to various targets.

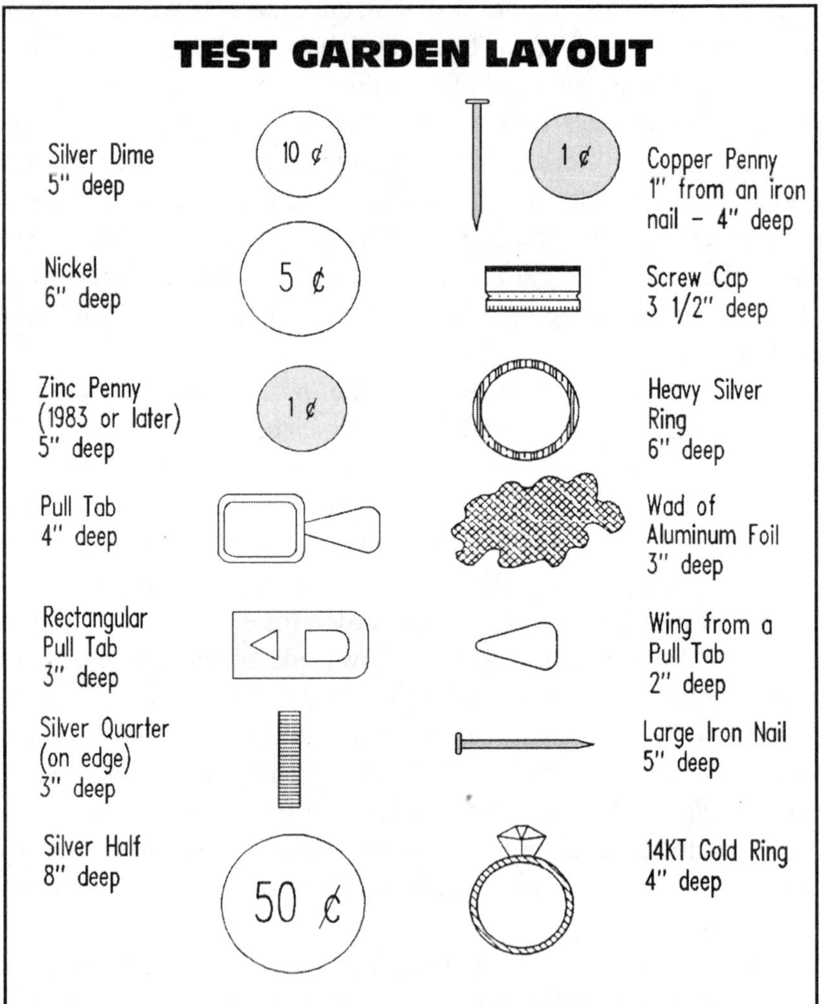

All of the objects in your test garden should be buried at least 18" apart and their locations marked with colored golf tees.

NOTE: Canadian treasure hunters report that modern, pre-1996 Canadian nickels, dimes and quarters are usually classified as rectangular pull tabs by the CZ-5/6/6a in the NORMAL mode. However, in the SALT mode (the CZ-20 is always in the SALT mode), the Canadian dimes and quarters are classified as zinc pennies.

If you have not searched for a certain type of target recently (you may have given up coin hunting during the hot summer months due to the dry ground conditions), a few minutes spent practicing in the test garden will familiarize you with the detector's response and will keep you from wasting valuable time in the field.

As ground conditions change, the target response can vary greatly. Some of the better finds made by my wife and myself came from digging signals that sounded questionable; however, due to the time spent in the test garden before we left the house, we knew they were "keepers."

Another effective use of the test garden is to practice and improve your pinpointing skills. Even though the PINPOINT mode on the Quicksilver is one of the easiest and most accurate of any available today, many users have reported that they have difficulty pinpointing targets in the ground. While accurate pinpointing reduces the amount of time it takes to recover a target, the most important reason this skill should be mastered is to prevent damage to the ground during the recovery phase by limiting the size of the hole that must be made. Sweep the search coil across an item in your test garden, and once you receive a clear signal, move the coil off to the side of where the target is. Press the PINPOINT button and slowly move the coil back and forth over the target area. Listen to the audio response (and watch the meter

NOTE: The depth indication on the meter has been calibrated to provide accurate values on coin-sized objects. Larger objects will appear to be shallower than they really are based on the meter indication they produce. Despite this "error" in depth reading, you will still be able to accurately center the coil over the center of the target regardless of its size by using the technique described.

indication on the CZ-5/6/6a). When the pitch reaches its highest level and the meter indicates the SHALLOWEST depth, stop moving the coil and release the PINPOINT button. At this point, the golf tee should be directly beneath the center of the coil. If not, repeat the process until you are able to consistently center the coil over the golf tee marking the target's location.

Remember, the time you spend perfecting your pinpointing technique will pay dividends in the field by enabling you to recover targets more quickly and with less damage to the ground.

Building and using a test garden on a regular basis will help you become more proficient with your metal detector in the field. It will also improve the quantity and quality of your finds. Remember, air tests can be deceiving; the only way to practice and improve your skill is to have targets buried in the type of soil you search routinely.

CHAPTER 3
COIN HUNTING

These exceptional coins from the 1800s are some of the coins found by Larry Hayes during his first four months of coin hunting home sites with a CZ-5.

Coin hunting is probably the most popular form of treasure hunting. After all, coins can be found virtually anywhere people have been.

The most obvious areas for finding coins are those sites that are still in use on a regular basis. These can include school yards, parks, ball fields, picnic areas, roadside rest areas—the list is almost endless. While most of the coins found at these locations are relatively new and worth only their face value, the experience you gain interpreting what your detector is telling you will improve your overall success in the field. I make it a point to visit the local playgrounds and parks on a regular basis just to sharpen the skills I use when searching the sites that contain older coins and other valuables. Often the targets that are at the edge of the detector's detection depth are the coins that other treasure hunters have missed due to the marginal signal that they produce.

According to a number of seasoned treasure hunters from across the country, the CZ truly excels in finding old, deeply buried coins that previous hunters have missed due to the limitations of their equipment. For example, since acquiring his CZ-5, Larry Hayes, a treasure hunter with only four months of experience, has recovered such coins as an 1812 large cent, an 1852 Seated Liberty dime and three Barber halves from his home town in upstate New York. It should be noted that all of these finds came from areas that had been heavily hunted by dozens of other treasure hunters for nearly 20 years. Dave Sponenberg Sr., who introduced Larry to treasure hunting, has also begun using a CZ exclusively in his searches. Dave, who has over 15 years of experience under his belt, reports that both the quantity and quality of his finds have improved since switching to the CZ-6. Specializing in searching sites that others (including himself) had written off as being "worked out," he has been able to add literally dozens of early date copper and silver coins to his collection, including a 1782 Spanish reale.

A common observation made by Larry, Dave and other successful coin hunters from around the country has been that most coin hunters frequent sites like modern parks, schools and playgrounds, yet wonder why they are not finding any older coins. To find the early date coins we are all searching for, you need to visit those sites that were in use during the time when the coins were in circulation.

One of the most-often overlooked places for finding old coins and artifacts dating as far back as the 1600s are old home sites. These can either be currently occupied or long-since abandoned. The advantage of hunting home sites over most other areas is two-fold. First, fearing rejection, most treasure hunters tend to avoid knocking on the door of a house and asking the owner if they can detect the yard. As a result, *you* will often be the first person to search a yard if you are able to obtain permission. Secondly, yards generally contain very little trash, which makes searching them much easier than the local public park or picnic ground that is littered with pull tabs and screw caps.

If areas such as local parks and schools make good choices for finding recently lost coins, then similar types of sites that were in use prior to 1900 would obviously be well suited for the coin hunter who is looking for key-date coins. Surprisingly, less than 25 percent of all such sites have been visited by metal detectorists, and most of those that have been hunted were not covered in an

Roy Gouge sits behind the hundreds of finds he made with his wife, Phyllis, in Massachusetts. The Gouges use CZs almost exclusively in hard-hunted areas already searched by hundreds of treasure hunters.

effective manner or with high-quality equipment. As a result, they still contain a large number of good targets.

Examples of these types of sites that have proven productive in the past include picnic groves, church revival and meeting sites, fairgrounds, organized work camps (such as the WPA), rest areas on roads long since bypassed by super-highways, old motels, sledding and ski areas and swimming beaches.

County atlases for many areas of the country were published from the mid-1800s through the 1920s. If an atlas was published for your county (or surrounding county), try to obtain a copy of it either from your local library or historical society. These atlases will clearly show details of where parks, picnic groves, race tracks and other recreational sites were once located. Often the only additional effort required to locate the site is to transcribe the coordinates onto a current map and drive directly to it.

If you are unable to find an atlas or if one was not made for your area, try checking for a detailed map of the area. If your library does not have any maps that meet your needs, you should contact K.B. Slocum Books. This is a source that I have used frequently over the years when I have needed a map for an area and could not find one though any other source. They have an extensive inventory of maps and other reference materials that dates back to the 1700s for virtually any part of the United States. To find out what information they have on your specific area, write them at K.B. Slocum Books, P.O. Box 10998, Austin, TX 78766 or call (800) 521-4451 and request a copy of their latest catalog.

One of the easiest methods you can use to locate productive sites is to ask other treasure hunters you know or meet what areas they consider to be worked out. Often these sites still contain a large number of old coins and other valuables while most of the trash and modern coins have been removed. For example, Roy Gouge and his wife, veteran treasure hunters from Massachusetts, have averaged over 5,000 coins a year searching heavily worked sites near their home since switching to CZ-6s two years ago. Included in these are two gold coins—a 1917 British sovereign and an 1881 U.S. $5 gold piece. This clearly shows why researching sites others have given up on can be extremely rewarding when the proper techniques and equipment are used.

Join your local treasure hunting club and ask members about sites that once produced well but are now avoided due to the pressure they received. Remember, as more sensitive equipment becomes available, once-productive sites are worth revisiting. The CZs are ideal for this purpose since their above-average detection depth and discrimination capabilities will allow you to detect coins that are beyond the reach of most other detectors currently on the market.

As discussed in Chapter 1, the AUTOTUNE search mode provides increased detection depth and wider ground coverage with each sweep. Because of this, many coin hunters (and other types of treasure hunters as well) choose the AUTOTUNE mode as their primary search mode. While the AUTOTUNE mode does offer several distinct advantages over the motion discrimination search mode, its inability to identify and differentiate good/bad targets poses a problem when hunting sites that contain a fair amount of trash. In order to receive the best of both modes, a number of successful coin hunters have developed an effective search technique. They search in AUTOTUNE and once a signal is received, they check the signal by moving the discrimination control into the "0" position with their thumb or forefinger. Since both the meter and audio target identification circuits are activated once the discrimination control is taken out of the AUTOTUNE setting, it becomes quite simple to determine if each signal is worth recovering.

Another trick to remember when using this technique is that if no signal is detected once the discrimination control is placed in the "0" position, the target is probably beyond the depth at which the CZs circuitry can accurately identify it. In these cases, remove a plug and recheck the target. A number of users have reported finding coins and other small items dating back to the Colonial period at depths exceeding 12 inches that produced faint yet clear signals in the AUTOTUNE mode, but were undetectable in the motion-discrimination mode.

Most productive coin-hunting sites were frequented by a large number of people, and unfortunately, they usually left behind as much trash as they did coins and other valuables. While the discrimination circuit of the CZ will allow you to eliminate most of the unwanted targets, areas that contain a high concentration of trash will be virtually unhuntable due to what is known as signal masking. If a coin is buried adjacent to a piece of trash, the detector may get a stronger signal from the trash, and as a result, reject the coin along with the trash. The 5-inch coil produced by Fisher has been designed to significantly reduce the possibility of signal masking. Since fewer targets can be under the coil at a given point in time, good targets will not be rejected due to a nearby trash target that presents a stronger signal.

The 5-inch coil also excels when searching near metal objects, such as chain link fences, bleachers, parking meters and even free-standing pay phones. When using the standard 8-inch coil, the detector will begin to respond to the metal in these objects whenever the coil gets within 6 to 10 inches of the large, metallic object. The 5-inch coil will let you get much closer, often within

Careful searching with a CZ-5 in several well-hunted Denver parks using the techniques described in this book produced these coins, rings and other artifacts over a 3-week period.

an inch or two, before the detector's circuitry senses it. This allows you to detect and recover targets in areas most treasure hunters have been unable to hunt in the past.

Overlapping each sweep is one tip that can't be over emphasized. Remember, the electrical field generated by the search coil is conical in shape, and while the coil may cover 8 inches at the surface, it will only be "seeing" an area of 2 or 3 inches in diameter at a depth of 8 inches. By overlapping each sweep by at least 25 to 30 percent (and preferably 50 percent) you will reduce the possibility of missing deeply buried targets. The failure of most coin hunters to overlap each sweep is probably the main reason older sites still contain a large number of valuable coins, even after being hunted by hundreds of people over the years.

As proof of this technique's value, my partner and I recently went to an older elementary school located in central Pennsylvania. After adjusting my CZ, I began working along the edge of an old sidewalk that led to the side of the school building. Overlapping each sweep by 50 percent, I was sure I would not miss any target my coil might pass over. In the time it took me to cover a 50-foot section of the sidewalk, my partner had made several passes across the large ball-field area. After an hour he came over and showed me a few recently lost coins he had found. Somewhat disappointed, he indicated that he was ready to leave and try somewhere else. His expression quickly changed when I showed him four Indian Head pennies, two Buffalo nickels, several wheat cents and a sterling silver charm I had recovered from the small area I had just searched. Needless to say, we stayed, and over the next several hours, the two of us recovered over 50 coins from a site my partner had been ready to give up on.

Dan Phillips, a coin hunter from central Ohio, has developed a technique that has greatly increased the quantity and quality of older coins recovered from local parks that others consider to be "worked out." After watching dozens of hunters hurrying across the open expanses of the parks and schools surrounding his home, Dan realized that these were the areas that probably held the greatest number of old coins and other items. Selecting a small park that had been heavily hunted for over 20 years at which to try out his idea, he brought along a roll of kite string and several 12-inch-long wooden dowel rods in addition to his CZ-6. Using the string and dowels, he marked an area 4 feet wide by 50 feet long. This helped him make sure he would thoroughly work the area he had marked. Overlapping each sweep by at least 50 percent, it took over an hour to hunt the area bounded by the string. The results—six recently lost coins, two Indian head pennies, six wheat cents, an 1898 Barber dime, a Buffalo nickels and a small silver ring. Dan has used this technique with similar results at sites throughout the tri-county area surrounding his home.

One of the most frustrating aspects about coin hunting that virtually every treasure hunter experiences is receiving a signal and then after several minutes of effort, finding themselves holding a piece of trash. While the CZ's discrimination circuit will reject most trash targets,some trash requires operator skill to avoid wasted effort.

One technique that is quite effective in determining when a questionable signal should be left in the ground rather than taking valuable search time recovering it is based on the signal's repeatability. If you receive a signal that gives a solid "beep" in one direction yet disappears when the coil is swept in the other direction, it is probably a piece of trash not worth recovering. To check this before continuing on, turn 90 degrees and recheck the area. If you now receive a signal in both directions, you have probably come across a good target buried next to a piece of trash. In this case, use the PINPOINT mode to separate the two targets and recover the good one. If you still only receive a signal in one direction (or possibly in neither direction), the target is trash and

These silver coins and jewelry items, dating back to the early 1800s, were found by Dave Sponenberg Sr. in 1994 with his CZ-6 by using many of the techniques described in this book.

should be ignored. This technique can save you a considerable amount of time in the field, and remember, each trash target that you are not forced to spend time recovering leaves you more time to find and recover valuable ones!

Learning proper target recovery techniques is important for several reasons. First, careless recovery of an older coin can quickly render it virtually worthless. Many coin hunters have spent a considerable amount of time finding sites to search yet are unwilling to take the time needed to recover a target without damaging it. A few years ago I was hunting an old park in

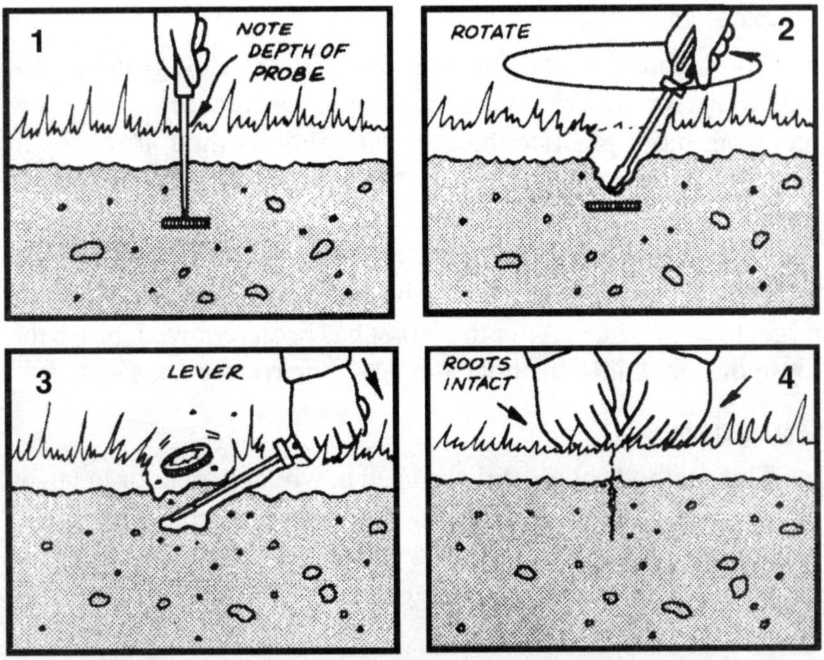

The Probe and Driver method of target recovery uses a probe to locate the target and a screw driver as a knife and a recovery tool.

Harrisburg, Pennsylvania and received a signal indicating a deeply buried target. Without thinking, I pressed my knife into the damp soil and cut a plug. Bending the plug back, I saw a silver coin lying in the bottom of the hole. The coin, an 1892 Barber dime, would have graded AU except for the three deep scratches my knife blade had put on the face of the coin. My carelessness had turned a $150 coin into a "filler piece" worth simply what the silver content would bring.

The second reason proper recovery techniques are so important is that many areas have been closed to metal detecting after damage was caused by coin hunters. By utilizing the correct recovery method for the ground conditions present when you search an area, you can recover literally dozens of targets without leaving any sign of having been there. Not only will this make the owners of the property happy, but other treasure hunters will not see your holes. I have often hunted an area simply because I could see signs of someone else who hunted it and had obviously recovered a number of targets.

Probe & Driver

This technique should be used in areas where the ground is not that moist or where cutting a plug is not acceptable. While requiring more practice than the plugging method, it results in virtually no damage to the lawn. After pinpointing the target, use a probe to locate the target. Next, insert an 8-inch screwdriver just above the target and rotate it slowly to spread the grass roots. Now insert the screwdriver just under the target at an angle and work the target to the surface. After the target has been removed, brush the loose dirt back into the hole and press the roots back together.

Plugging

This method should only be used in wooded areas or in lawns that are moist. Plugging in dry, hard ground can damage the grass, leaving yellow "dead spots" after you leave. After pinpointing the target, cut a plug three-quarters of the way around using a large hunting knife.

Cutting a complete plug will usually result in the grass dying

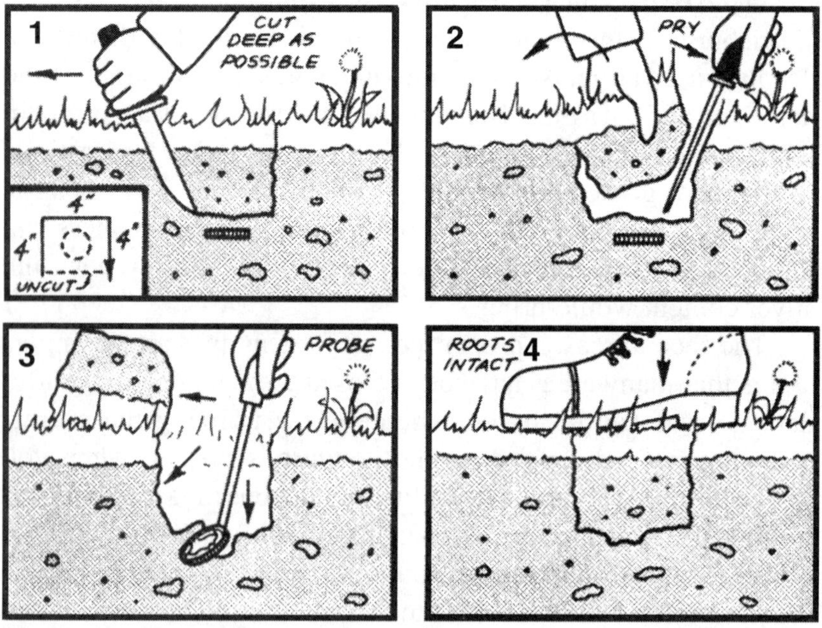

The Plugging method of target recovery, when appropriate, leaves virtually no evidence of digging. To avoid killing the grass, this method should not be used during the summer months.

within a few days time. Using the blade of the knife as a lever, fold back the plug. Sweep the search coil over the hole and the plug to determine where the target is located. Use a probe to locate the target and recover it.

NOTE: Do not use a knife with a folding blade as it may inadvertently close, resulting in injury.

A trick that many experienced coin hunters use to further help reduce the possibility of leaving any evidence of their having recovered a target is to bring along a 12-inch square piece of vinyl and put all of the loose dirt from the hole on it. Once the target has been recovered, simply pour the loose dirt back in the hole, fold the plug back in place, and step it down.

CHAPTER 4
RELIC HUNTING

Relic hunters have been searching for pieces of America's past since the 1940s when surplus Army mine detectors were the only tools available; however, relic hunting has only recently begun to grow in popularity. Movies such as *Gettysburg, North and South, The Deerslayer*, and *Last of the Mohicans* as well as the PBS mini-series *The Civil War* have generated a great deal of public interest in the tumultuous periods of this country's history.

Virtually every part of the country has been touched by various wars, including the American Revolution, French and Indian Wars, the War of 1812, the Civil War, the Indian campaigns of the 1800 and even the training and fortification activities associated with the first and second World Wars. Military relics left behind by the individuals that fought in these conflicts provide a bridge with the past for those that find them. When one recovers a minie ball from a Civil War battlefield, one can almost feel the battle that raged nearby some 130 years ago.

In addition to the historical nature of the relics being found by metal detectorists, the monetary value they hold is truly amazing. The following list provides a representation of what typically found artifacts are currently worth:

- Revolutionary War pewter button: $10 to $300+
- Civil War minie balls: $1 to $300 each
- Confederate belt buckle: $2,000 to $6,000+
- Confederate uniform button: $25 to $2,000+
- Civil War artillery shell: $100 to $1,000+
- World War I insignia: $5 to $100
- World War II post exchange token: $2 to $25

Before discussing some of the techniques you can use to improve your success rate in the field, a brief overview of some of the laws that restrict where relic hunting is allowed is warranted.

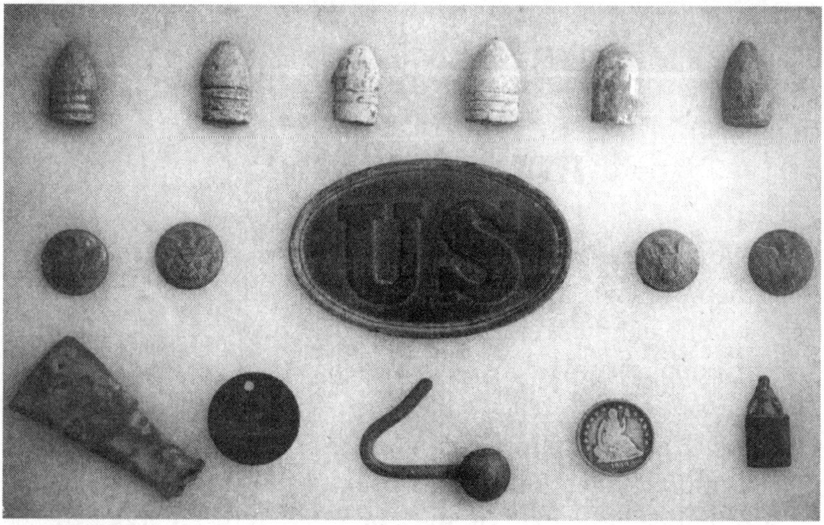

Civil War relics like these from the Battle of Atlanta can still be found on private property with a little research and a deep-seeking metal detector such as one of the QuickSilver models.

In the early 1900s, the federal officials recognized that historical artifacts were being collected by citizens at an increasing rate. As a result, the story that these artifacts could tell about life in a specific area was lost for future historians and archeologists. To prevent this loss of historical resourses, the Antiquities Act of 1906 was passed, which afforded protection to artifacts on government-owned lands. Despite the existence of this law, very little enforcement action was taken to protect artifacts, and as a result, removal of artifacts continued at an alarming rate. In 1979, the government passed another law known as the Archeological Resource Protection Act or ARPA. This law stipulates stiff fines and/or jail sentences for anyone convicted of removing artifacts from federally owned or Indian lands. This includes ALL national parks (such as most major Civil War battlefields), national seashore and any area under the jurisdiction of an American Indian tribe.

No longer ignoring even minor infractions of this law, federal agencies such as the National Park Service have begun prosecuting individuals caught removing artifacts (especially individuals with metal detectors) with a high conviction rate. Considering that fines can reach $250,000 and jail sentences can be up to 2 years, relic hunters should be extremely careful where they search for artifacts before they find themselves on the wrong side of the law.

In addition to the restrictions that exist on federal and Indian lands, other areas that may have restrictions on the use of metal detectors include the following:

- State Parks
- State Historical Sites (including areas around road-side markers)
- Sites designated as being on the National Historic Register
- Local parks that commemorate an historic event (battle, campsite, ferry site, etc.)

It is strongly advised that BEFORE you start hunting any public area, you check with the agency responsible for it and determine if metal detecting is allowed. With the large number of violations that have occurred in recent years, ignorance is no longer a valid excuse if you are caught hunting in an area where metal detecting is prohibited.

Before you give up on the idea of searching for relics, keep in mind that many of the smaller skirmish sites, campsites, or training grounds are on private property and can be searched with permission from the current land owner. Experts estimate that over 75 percent of all relics are on private property. With these statistics, relic hunters should never even have to think about risking possible arrest by searching areas that are designated as off-limits.

While most relic hunters are searching for military relics, there are numerous other types of artifacts that one can recover, like old home sites, railroad sidings and mining camps—virtually anywhere people have lived and worked. Artifacts being recovered by hunters cover a wide spectrum, ranging from personal items such as keys, silverware, pet I.D. tags and buttons to tools, farm implements and even weapons, such as pistols, rifles and swords. Many of these items are highly valued by today's collectors and can be quickly sold for a nice profit.

All of the QuickSilver models make excellent relic detectors because of their detection depth and accuracy of discrimination; however, the CZ-6, 6a, and 20 are preferred by most of the avid relic hunters around the country because of the water-resistant control housing of the CZ-6/6a and the waterproof design of the CZ-20's housing. Often relic hunters find themselves a mile or more from their car searching a productive site when an unexpected rain shower appears. The control housings on these models keep the electronics from being damaged by rain, and if you don't mind getting wet yourself, you can continue hunting and recovering targets.

Discussions with a number of successful relic hunters across the country revealed two distinctly different yet equally effective methods being used to hunt for relics with a QuickSilver. The first is similar to the AUTOTUNE method described in the previous chapter. Some relic hunters will search in the AUTOTUNE mode for the nominal increase in overall detection depth it produces, and then once a target is detected, they move the discrimination knob to the "0" position, activating the audio and visual target identification circuitry to determine if a target is worth recovering. Other relic hunters opt to search in the discrimination mode with the control set at the "0" position. Whenever a target is detected, the target identification circuitry immediately provides the hunter with a probable target I.D. without having to move the discrimination knob. Hunters that use this method feel that the increased efficiency and ease of operation outweighs the slight loss of detection depth that may occur. You should try both methods and use the one that best suits your individual hunting style.

Unlike the typical coin hunter, relic hunters normally search locations that contain very few of the trash items that coin hunters face, such as pull tabs, screw caps and tinfoil. However, this does not mean that the only signals one encounters will be valuable relics. Trash targets (square nails, strands of barbed wire, pieces of old cans and other rusted iron) frequently outnumber the relics you are hoping to find at many sites.

The most important point that relic hunters need to remember to avoid missing any relics is that discrimination should be used SPARINGLY. Unless the area is heavily littered with old nails or pieces of barbed wire, the discrimination control on the CZ should not be set above "0." In essence, your detector will be operating in an all-metal mode with every target being identified audibly and visually (Note—the meter is only on the CZ-5/6/6a). If you find yourself in an area that contains a high concentration of small, iron trash targets, set the discrimination control at "1" to eliminate signals from those objects. Due to the wide variation in size, shape and composition of relics, it is virtually impossible to use the segments on the meter to accurately identify all targets before they are recovered. As a result, most relic hunters simply use the discrimination circuit to reject small iron objects and recover any target that produces a medium or high audio signal (and registers in the non-iron area of the meter on the CZ-5/6/6a). Larger iron objects, such as plow points, ax heads, artillery shell fragments and gun parts, will still produce a signal; however, the signal will not be as smooth as that produced by a coin or a button. While some relic hunters would prefer having a detector that eliminated all iron, they need to remember that some of the better finds one can make consist of iron. By combining the excellent performance

TARGET	AUDIO SIGNAL
Revolutionary War musket ball	Medium
.36 cal. Civil War pistol bullet	Medium
.577 cal. minie Ball	Medium
.69 cal. minie Ball	High
Revolutionary War pewter button	Medium
Civil War cuff button	Medium
Civil War coat button	Medium
Sword scabbard tip (brass)	Medium
Artillery cap insignia (brass)	Medium
Confederate belt plate	High
Union eagle breast plate	High
WW II officer's hat emblem	High
Civil War artillery shell (Parrot)	Medium
Brass skeleton key	High
Sterling silver fork	High
Brass pet identification tag	Medium

These are typical audio signals from "good" targets.

This Colonial-period silver brooch and three Revolutionary War musket balls were found in a plowed field by Dave Sponenberg Sr. with a CZ-6 in upstate New York.

of the CZ with some operator skill in interpreting what the detector is telling you, you will find that the quality of your relic finds will increase dramatically.

The table on the perevious page shows the type of signal produced by a representative sample of relics that hunters typically recover. As you can see, ignoring only the low-tone (iron) signals will ensure that you do not inadvertently leave a valuable target in the ground.

If you get a signal that does not lock on to a specific segment of the meter (or gives mixed audio tones), switch to the pinpoint mode and locate the exact center of the target or targets. Release the button and sweep the coil back over the target to obtain an accurate reading. Often two targets in close proximity to one another will cause this problem. The CZs do an excellent job in alerting the operator to the fact that there is more than one target under the coil, which allows the operator to determine if either is worth recovering. On many other target-I.D. detectors, multiple targets will produce a single, incorrect reading, resulting in the relic hunter leaving a potentially valuable target in the ground for the next person to find.

Large, rusted objects may produce what many relic hunters consider to be false signals. This is what happens: as the coil passes over the object, the discrimination circuit accurately identifies it as trash. No audio/visual signal is produced (if searching in the discrimination mode with the control set at "1" or higher). Then, as the coil reaches the edge of the object, the CZ's circuitry resets itself. If the object is large or very close to the surface, you will receive a signal that sounds and registers like that of a good target. The way you can avoid digging these types of targets is to sweep the coil across the target from another direction. If you receive a signal in a different place depending on which direction you sweep the coil, the target is probably trash. As a final check, switch to the pinpoint mode and check the target area. If it is a large, single object, it's more than likely a piece of rusted metal and can be ignored.

NOTE : If you are looking for iron artifacts, such as axe heads, cooking implements or knives, the discrimination control should be left in the "0" position and all of the larger iron signals recovered.

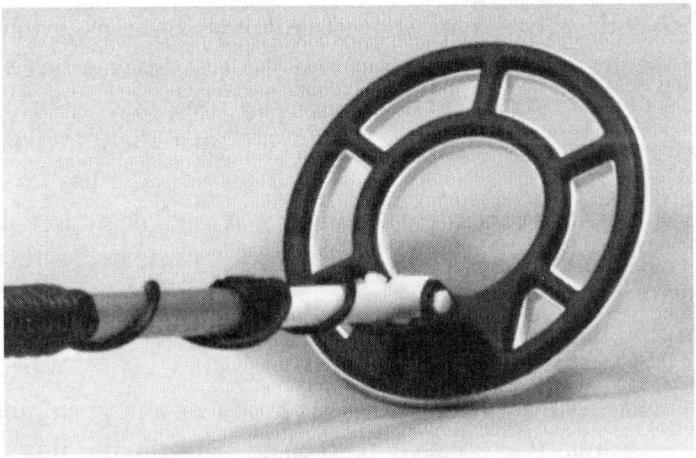

Coil covers are made for every model of Fisher detector. Made of lightweight but rugged plastic, they protect the underside of the search coil from nicks and scratches.

Selection of the correct search coil size for a specific situation is another factor that can greatly affect your overall success in the field. Many relic hunters feel that using the 10 1/2-inch coil is necessary to reach deeply buried relics that others have missed. While this may be true in some cases, using the larger coil may actually cause you to miss relics that would have been found with a smaller coil.

The 10 1/2-inch search coil gives an average 20-percent increase in detection depth on coin-sized or larger objects; however, it has two drawbacks when compared to an 8-inch or 5-inch coil. First, in an area where there may be more than one target under the coil at the same time, a deeply buried good target may be masked by a trash target just under the surface. Many campsites and old homes have a high concentration of small iron targets, such as boot tacks, nails, wire, rusted cans and wagon parts, which can mask good signals, particularly from targets that are low in conductivity, such as gold and lead. Also, the physical size of the larger coil prevents you from getting into areas like rock piles and briar patches, which often hold relics others have missed. The 10 1/2-inch coil is recommended for searching open areas containing widely scattered relics or when you want to cover a large area quickly to determine where the activity took place. Once you locate a promising area, you will be better off switching to one of the smaller coils.

The QuickSilvers provide above-average detection depth with the standard 8-inch coil, which is why most successful relic hunters prefer using this coil in their searches. In fact, the 5-inch coil provides detection depth equivalent to that of many other detectors equipped with an 8-inch coil, yet allows for much better target separation. The 5-inch coil has produced a large number of relics from areas that others had given up on due to the amount of trash that littered the area.

If you are using the 8- or 10 1/2-inch coils, you should purchase the solid coil cover, which keeps the search coil from getting snagged on underbrush as you hunt. Coil covers featuring open and closed designs for these coils can be purchased from any Fisher dealer.

CHAPTER 5
BEACH AND WATER HUNTING

Searching for valuables lost along beaches, both in-use and abandoned, has become the fastest-growing facet of treasure hunting over the last few years. The reason for this growth is well-founded, after all, beaches probably hold more coins and gold or silver jewelry than any other area accessible to the recreational treasure hunter. Many water hunters have made finds that make even professional treasure hunters take notice, including rare Spanish gold coins, rings worth more than $20,000 and historic artifacts spanning hundreds of years.

While all four QuickSilver models make top-notch beach hunting detectors, the introduction of the CZ-20 has given beach and water hunters performance that was previously unavailable in a fully waterproof detector.

There are two different types of beaches (fresh and salt water), with each having its own specific problems and corresponding techniques that should be used in order to maximize your success. Due to the differences between the two types of beaches, this chapter is divided into two sections aptly titled "Fresh Water" and "Salt Water." It is worth reading both sections even if you only plan on hunting one type of beach since several of the operating tips are only given once yet apply equally to both types of beaches.

Fresh Water

Freshwater swimming sites are by far the easiest of the two types to hunt for several reasons. First, with the exception of some river sites, there will be no current action to move valuables from the spot at which they were originally lost. Second, most fresh-water beaches are smaller than ocean beaches, and as a result, valuables tend to be more concentrated and easier to find. And finally, unlike saltwater beaches, freshwater sites tend to contain very little mineralization, which often results in a loss of overall detection depth. Despite the apparent advantages freshwater sites have over saltwater sites, there are still several problems that freshwater hunters must face.

The first is the sheer number of targets that even a moderately popular site will contain, both good and bad. On saltwater beaches, iron objects such as nails, bobby pins and hair berrettes rust away in a short period of time. Unfortunately, this is not the case at freshwater sites. These trash targets will often remain exactly where they were lost for years, and this concentration of ferrous trash can adversely affect your success.

By design, the QuickSilver series rejects virtually all small, rusted iron with the discrimination control set at "1." Most water hunters either hunt with the discrimination control at "1" to reject the iron signals or at "0" and then use the audio/visual indication to determine if the target is worth recovering (low tone = iron). The advantage of hunting with the control in the "0" position is that you can determine if there is more than one target beneath the loop: i.e., a coin or ring next to a nail. The AUTOTUNE mode can be used for better coverage and detection depth; however, if the beach contains a large number of ferrous targets, the effort required to check each and every signal by switching to the discrimination mode will not be an efficient use of your time. You need to decide how much time you have, and if the extra effort can be justified on a case-by-case basis.

In addition to the items mentioned above, non-ferrous trash such as tinfoil, pull tabs and screw caps will also be found in surprising quantities. Despite the CZ's excellent discrimination capabilities, beach and water hunters are best advised to keep the discrimination control set no higher than "1" to avoid missing any jewelry items. As shown in the chart on the next page, gold rings can register in six of the seven categories depicted on the CZ-6/6a and CZ-5 meter. While hunting at a low discrimination setting will have you digging more trash, finding a petite engagement ring with a 1.5-karat diamond in it will more than compensate you for the additional effort.

As a matter of fact, some of my best finds have come from beaches that were previously searched by treasure hunters who got tired of digging pull tabs and increased their discrimination to the point where the tabs (and most gold rings) were rejected. Remember, the reason you are hunting a beach is to find jewelry, not recently lost coins. If you don't want to pass up that $2,000 ring, you will have to dig a fair amount of trash, due to the wide range of readings jewelry can produce.

It's a sad-but-true fact that most gold rings look just like a piece of foil or a pull tab to a metal detector, as shown by the QuickSilver meter readings of a sampling of 161 gold rings. When beach hunting, it is best to recover all targets other than those that register as iron to avoid missing a valuable piece of jewelry.

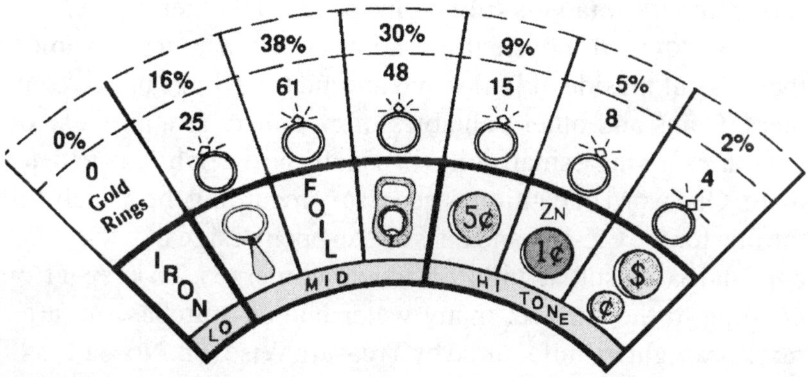

The lower row of graphics shows the CZ-6a QuickSilver audio response chart as it appears on the CZ-6a control panel. The rings indicate the number and percentages of gold rings that fell into each category when a random sample of 161 rings was identified by the CZ-6a.

An interesting "quirk" in the CZ circuitry is evident when the coil passes over multiple targets at the same time. On a number of occasions, I've received both a high and medium tone when sweeping back and forth across what I thought was a single target. In each case, I found a coin and a pull tab, or even a coin and a piece of jewelry in the scoop. This shows that the CZ will identify each target under the coil, thereby allowing the operator to decide whether or not the target is worth recovering. Most other detectors will combine both signals and produce an indication that does not accurately identify either one. So, if you receive a combination of signals, it's best to check and see what the area might hold. You could be walking away from an exceptional find lying next to a piece of trash.

While the standard 8-inch search coil is well suited for most beach and water-hunting applications, the other coils offered by Fisher can greatly improve your overall success under the right conditions. Discussions with both Fisher factory personnel and successful water hunters across the country indicate that while the 5-inch coil is probably one of the most useful accessories available to the CZ user, it is also one of the most under-utilized and overlooked options in the Fisher line. Providing only slightly less detection depth than the 8-inch coil, it does an excellent job in separating trash targets from coins, rings and other items.

In addition, its physical size allows users to hunt areas among the rocks alongside of jetties and around the pilings of docks and piers. Coins and other valuables often collect around these obstructions. Using a small coil will usually be the only way to detect them. One word of caution is necessary regarding the 5-inch coil: the one for the CZ-5, 6 and 6a is not an open design like the 8-inch coil, and as a result, it will float when submerged. To keep it from popping to the surface, many water hunters purchase an aftermarket weight manufactured by Treasure Wise (Ph. No. 414-542-7772). It is available from most Fisher dealers, and it simply attaches to the shaft. Another simple, yet effective trick is to fill a sock with sand and tie it around the shaft. The 5-inch coil on the CZ-20 is negatively buoyant and does not require any additional weight to keep it submerged.

The 10 1/2-inch coil is more suited to beach hunters who search the dry sand down to the surf line. In addition to providing approximately 20 percent more depth on coin-sized objects than the standard 8-inch coil, it also allows you to cover more ground with each sweep. While some water hunters prefer the larger coil for the improved detection depth it provides, there are several difficulties associated with its use in the water that one should consider before switching from the standard coil. First, the larger coil will be harder to sweep through the water due to the increased drag its size produces. Second, pinpointing is more difficult, and if you are searching an area with waves or a strong current, it may be difficult to remain in one place long enough to recover a deeply buried target, especially if you are not able to pinpoint it accurately. And finally, the larger coil is more susceptible to "target masking," which, if the beach contains a fair number of targets, may cause you to miss a valuable piece of jewelry or rare coin.

A highly effective beach- and water-hunting technique that is unique to the QuickSilver series involves the audio target-identification system. By setting the discrimination control at "1," only those targets that produce a medium or high tone will be heard through the headphones or built-in speaker on the CZ-5. Now, if you ignore any high tone (remember that all coins including nickels will produce a high-pitched tone) and only recover those targets that produce a medium tone, the amount of jewelry you recover in a given amount of time will increase significantly. Looking at the gold-ring chart on page 49, you can see that only 27 of 161 rings would have been missed if the high tones were ignored.

Now I'm sure many of you are wondering why I would recommend passing up signals that are obviously coins; however, let me first say that this technique should not be used at all times. If you only have a short amount of time to hunt a site, then wouldn't you be better off recovering a few gold rings rather than $10 in coins and possibly one ring? For example, a few years ago (before I had a CZ), my partner and I hunted a popular beach in Pennsylvania. Over a two-day period I recovered almost 600 coins and 8 gold rings from a small portion of the beach. If I had been able to ignore all of the coins, just imagine how many more rings I would have found!

Another time this technique will help you improve your jewelry-recovery rate is when hunting a beach that only allows metal detecting after the beach closes for the season. Often there will be 10 to 20 hunters searching the beach the day after it closes. If you leave all the coins for the other hunters and simply focus on medium-tone targets, you will probably be able to find more gold than all of the other hunters combined. Then, after the initial rush fades, you can come back and recover any coins that the other hunters might have missed.

Several divers I spoke to indicated that having an accurate target-identification circuit on a fully waterproof detector like the CZ-20 has allowed them to work areas that had been too trashy to hunt in the past with other detectors. Many of these divers report finding literally hundreds of old coins and other artifacts while hunting around old ferry landings, wharves and once-popular beaches. The CZ-20 is truly a versatile detector since it can be used for land hunting and wading and then collapsed small enough for diving, without the need for any additional parts or conversion kits.

Jerry Denton found this handful of old coins and 17 gold rings—including one with a .63-karat diamond worth over $1,000—while diving with a CZ-20 in a popular swimming lake over a three-day period in upstate New York .

Salt Water

Many of the tips and techniques described in the section on fresh water will be applicable to hunting saltwater sites as well. Hunting the dry-sand region of ocean beaches is very similar to hunting an inland beach; however, once you get down to the wet sand or venture into the surf, things become a little more difficult.

Historically, VLF-type detectors have not been able to function at peak performance levels in and around salt water. This is due to the way salt affects the circuitry, often producing falsing and chattering even at extremely low sensitivity settings. Typically, the only type of detector that could work this area effectively was a pulse-induction unit. While these detectors were able to find deeply buried targets, they lacked any type of discrimination, and as a result, treasure hunters often spent a considerable amount of time recovering trash targets, including nails, wire and tinfoil. The combination of the dual operating frequencies and salt mode effectively maintains the CZ's exceptional detection depth in and around salt water. This is especially useful when searching the wet sand area near the surf line.

When hunting the dry-sand area, the SALT/NORMAL toggle switch on the CZ-5, 6 and 6a should be left in the NORMAL position. As you approach the wet-sand region, simply place the toggle switch in the SALT position, reground balance the detector and continue hunting on into the surf. No longer will you experience the chattering found on other VLF units, yet the discrimination capabilities will allow you to ignore virtually all of the ferrous trash commonly found at popular swimming sites (the SALT circuitry is always active on the CZ-20). Just remember the discussion on setting the sensitivity control given in Chapter 1. If you set the sensitivity control too high for the specific conditions at the site you are searching, you can still get excessive falsing and chattering even in the SALT mode. Most saltwater hunters keep the sensitivity control set at "5" or below and still are able to locate valuables at depths exceeding 10 inches. If you receive more than one or two false signals in every few sweeps, turning your sensitivity control down slightly should eliminate the problem.

Using the 10 1/2-inch coil on saltwater beaches can also result in false signals. This is caused by the adverse effect of the combination of mineralized black sand and salt water, which is magnified by the larger coil. Unless your beach contains very little mineralization (such as the beaches of South Florida or the Gulf Coast), you will find that the standard 8-inch coil will provide above-average detection depth without the falsing produced by the larger coil.

NOTE: The 10 1/2-inch coil is advantageous when searching dry-sand areas because of the increased area you can cover with each sweep as well as the additional detection depth realized.

Salt water also tends to affect the audio and visual target-I.D.

readings produced by certain types of targets. For example, aluminum objects such as pull tabs and screw caps corrode quite readily when exposed to the salt and sand environment found on ocean beaches. This corrosion will often cause these common trash targets to register as a nickel; for example, they might produce a high audio tone and register in the upper portion of the meter. And zinc pennies that have been exposed to the salt for more than a few weeks will often register as a rectangular pull tab. So, if you are an avid saltwater beach hunter and don't want to miss any gold rings or coins, you will need to dig all targets that produce a medium or high tone.

While the QuickSilver is capable of detecting targets deeper than that of most other detectors currently on the market, targets at the edge of the CZ's detection capabilities tend to result in meter indications that jump from one segment to another or produce mixed audio tones. The trick to determining if these questionable signals are worth recovering is to use the pinpoint mode and check the target's depth (either on the meter or based on signal strength). If the target is fairly deep, remove some sand and re-check it—the meter indication should lock-on, making it easy to determine whether or not the target is a "keeper." This technique will enable you to recover targets at depths that most treasure hunters would find hard to believe. For example, Larry Dunnagan, a South Carolina beach hunter with years of experience under his belt, frequently recovers coins and jewelry items at depths in excess of 10 to 12 inches from the densely packed black-sand beaches in his area.

With the release of the CZ-20, divers have begun looking at underwater treasure hunting with a whole new perspective. No longer forced to accept a mediocre detector in an underwater housing, divers are revisiting sites that were thought to be "cleaned out." Saltwater shipwrecks that could not be hunted previously due to the lack of discrimination on pulse detectors are producing artifacts at an impressive rate. For example, two divers based out of Rehoboth Beach, Delaware have been revisiting several well-known local shipwrecks with their CZ-20s. These shipwreck sites had been searched by literally hundreds of divers before them. Their finds in a four-month period have included a ships bell, six

Larry Dunagan, a South Carolina beach hunter, has recovered these rings, chains and pieces of jewelry (and more) with his CZ-6 and CZ-20.

port holes (two with the glass still intact), dozens of pieces of sterling silver tableware, brass fittings dating back to the early 1800s and personal items from crew members and passengers, such as coins, rings, watch fobs and belt or shoe buckles.

Some of the features on the CZ-20 that have received positive reviews by divers include the following:

- **Ground Balance Circuit:** Since mineralization and/or salt content can change significantly at sites even a short distance apart, having the ability to quickly adjust the ground-balance setting for the exact conditions at the site being searched with a control on the outside of the control housing ensures maximum detection depth is always obtained. Divers simply reach the bottom, press and hold the pinpoint button, lower the coil, and use the GROUND control to compensate for the mineralization/salt present.
- **Audio Boost Circuit:** The older, more valuable targets usually produce much fainter signals than those near the surface. On most underwater detectors, the signals from these types of targets are often difficult to hear over the bubbles coming from the regulator. The CZ-20's audio-boost circuit eliminates this

problem by keeping the volume of surface targets constant while amplifying the weaker signals that other divers have missed in the past.

- **Salt Mode:** Fisher's 1280-X Aquanaut was one of the first VLF-type detector useable in salt water. The submersible CZ-20, however, has advanced the technology a giant step. It performs in salt water every bit as good as a pulse induction

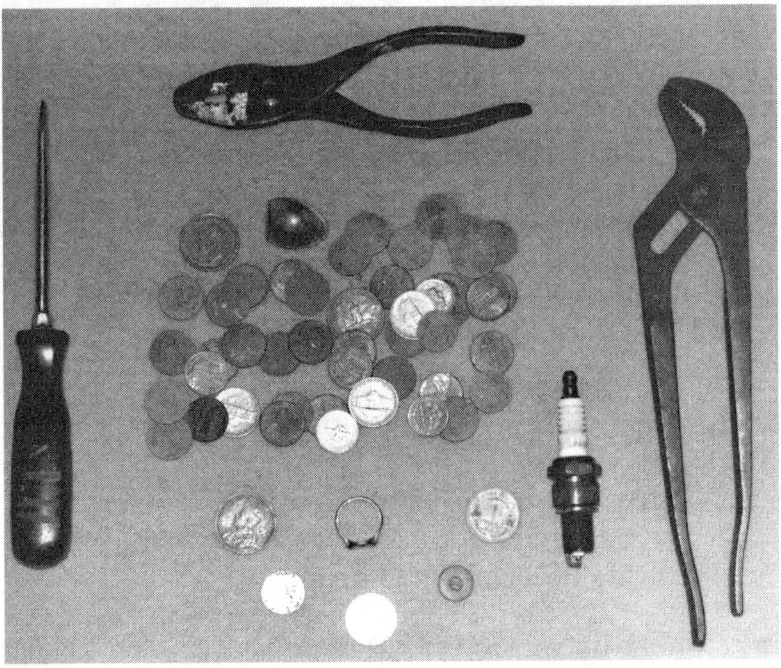

These tools, coins and rings were found by the author in 30 minutes while diving with a CZ-20 below a floating boat dock.

metal detector (until now, the choice for most professional underwater treasure hunters). Yet unlike a P.I., it has the ability to reject and/or identify many small items just like the very best land detectors. Underwater metal detecting is not limited to searching around old shipwrecks for artifacts. Many divers have found that areas such as boat docks, old ferry crossings, bridges and bodies of water that border old battle sites can be extremely productive. For example, a diver from Huntsville, Alabama has started searching the bottom of the Tennessee River near several well-known Civil War battlefields. Unlike the shoreline, which has been heavily hunted for more than 30 years, the river bottom has produced hundreds of artifacts, including bullets, buttons, artillery shells and even a Confederate sword. Many rivers, ponds or lakes located next to battlefields or campsites are potentially productive sites worth exploring with a CZ-20 (just make sure it is not on land where metal detecting is prohibited—see Chapter 4, Relic Hunting).

Beneath piers and floating boat docks are other areas that very few divers have even thought of searching. With the CZ-20's discrimination control set at "1," nails and other small pieces of iron frequently found in these locations will be rejected while the CZ-20 still responds to coins, jewelry and tools. Some of the best spots to search are boat docks that have gas pumps on them since boaters have a reason to take money from their pocket, and a fair amount of the change will wind up on the bottom waiting for someone with a metal detector to find it.

CHAPTER 6
ELECTRONIC PROSPECTING

The fastest-growing form of prospecting is nugget hunting with a metal detector. With the inherent stability and sensitivity of the QuickSilver metal detectors, more and more people have begun using them to search known gold-producing areas for nuggets—and many are quite successful in their quest! Most of the areas that were heavily worked in the 1800s still hold a small fortune in nuggets just waiting for someone with a metal detector to uncover them.

The tools needed for this type of treasure hunting include a quality metal detector, a small shovel, a magnet, a plastic gold pan and a container for your finds. The gold pan needs to be made out of plastic since you will be passing the search coil over it to determine when you have removed the target from its resting place.

There are several different areas you can search for nuggets: in or near a stream, in the dry areas common to the Southwest, in the tailing piles left by the bucket dredges that operated in the 1800s and early 1900s and around old placer mines. In all cases, the area should be a known gold-producing area.

Once you have located an area that has the potential for producing nuggets, turn on the detector and ground balance it as accurately as possible for the mineralization present. Increase the sensitivity to the highest setting that does not produce excessive falsing and chattering. Remember to recheck the ground-balance setting whenever you adjust the sensitivity control. When sweeping the search coil, be very methodical, and don't try to cover several acres in one afternoon. To search an area effectively, you will need to overlap each sweep by at least 40 to 50 percent. The reason for this is that the search coil sends out a cone-shaped signal. While an 8-inch coil will "see" the entire area underneath it at the surface, it will only be seeing a 2- or 3-inch circle at a depth of 7 or 8 inches. As you can see, without overlapping each sweep, it is quite easy to miss a small, deeply buried nugget. Remember, it is better to thoroughly cover a small area than rush and search a large area haphazardly.

An important tip to remember when prospecting is to frequently check the ground-balance setting. In many areas, the amount of mineralization can change dramatically within a short distance, and if an adjustment is not made, your detection depth will be significantly reduced. I persnally recheck the ground-balance setting on my detector every 5 minutes or when I have moved 30 to 40 feet from the point at which I last checked it. While this may seem excessive, it helps me make sure I am maximizing the detector's performance capabilities.

Discrimination is a feature that is much more important if your metal detector will be used for coin hunting in trash-filled parks rather than electronic prospecting. Most of the gold nuggets that you will be searching for will be quite small and extremely easy to reject if an excessive amount of discrimination is used. In most areas, you will want to hunt in the AUTOTUNE (all-metal) mode and recover every target located by the detector.

Once you have located a target, begin placing small amounts of dirt into the gold pan. After each shovel of dirt is removed, check the hole as well as the dirt in the pan. If the signal is still in the hole and no signal was detected in the pan, the dirt should be dumped out and the process continued until the target has been removed from the ground.

Unless the target is readily obvious, you should run the magnet through the dirt, which will quickly remove any small pieces of rusted metal. If the target is still in the pan, you can either pan it down if you are near a source of water, take the dirt out by the handful and run it over the coil or place it in a 5-gallon bucket and recover the targets at a later date, which is the method preferred by many successful prospectors.

Again, always check the hole for multiple targets—you'd hate to leave a hole after recovering a rusted nail only to have your partner come along and find a nice gold nugget a few inches deeper.

Another important point to remember when nugget hunting is to not get discouraged if you find a number of worthless targets between nuggets (and this will happen). In most known gold-producing areas, there have been people searching for gold for more than 100 years, and most of them will have left behind trash for you to find. If you are persistent though, your searches will become more and more successful as you gain experience and learn what your QuickSilver is telling you.

While hunting in the AUTOTUNE mode is the preferred method when prospecting, there are times when the CZ's discrimination circuit can actually help you recover more gold than someone who is hunting strictly in the all-metal mode. Often you may find yourself in an area that contains a large amount of ferrous trash such as nails, rusted wire or pieces of tin cans. The following technique will allow you to ignore most of these targets while still recovering virtually all of the gold that your detector's coil passes over.

NOTE: Frequently more than one nugget will be found in the same area, so be sure to recheck the hole and the dirt in the pan before dumping it out.

If you receive an extremely faint signal, it is best to recover it, as the discrimination circuit may not be able to accurately identify the target's composition. These signals can either be from a very small metallic object (such as a tiny gold nugget or a lead sliver from a bullet) or a deeply buried object at the fringe of the detector's detection capabilities. If you remove an inch or two from the area and the signal strength increases, this tells you the target is deeply buried. Continue removing dirt until the signal is strong enough to allow the discrimination circuitry to identify the target.

On signals that are strong enough to be identified, move the discrimination knob with your thumb from the AUTOTUNE position to the first discrimination setting labeled "0" (use your free hand if you have hipmounted the control housing). Sweep the center of the coil over the center of the target several times while watching the meter and listening to the signal produced. If it consistently indicates "IRON" and the audio signal is a low tone, give up on the target and continue searching. If you receive either an inconsistent meter/audio response or one that registers in any category other than IRON, the target is worth recovering. Again, only use the discrimination circuit to reject larger pieces of iron—

any small target should be investigated to avoid possibly missing a gold nugget!

Another piece of equipment that you might want to buy is a smaller search coil. The 5-inch Fisher search coil is more sensitive to smaller objects, and the detection depth is only slightly less than the standard 8-inch coil that is used for most applications. Another benefit of using the smaller coil is its ability to get into narrow crevices and under rock outcroppings that would be inaccessible to the larger coil. Many times, the nuggets you are hoping to find will be trapped in these locations and only the smaller coil will be able to detect them.

CZs can also be used to locate potentially productive spots to pan, sluice or dredge when nuggets are either too small or deep to be picked up directly by the detector.

Again, as in nugget hunting, the search mode one should use is the AUTOTUNE mode. The detector should be ground balanced using one of the two methods described earlier, and the audio threshold increased slightly above the level at which you would normally search. This is done by turning the sensitivity control (which serves a dual purpose as described in Chapter 1) in the clockwise direction once the ground-balance adjustment has been made. At this point, simply search along the stream bed, dry wash or mine site as you normally would, paying particular attention to the threshold signal.

In addition to the sharp, well-defined response from a metallic object, be alert for black-sand conditions. Pockets of black sand, which often contain gold flakes and nuggets, can produce either a positive or negative effect on the detector's audio signal depending on their mineral content. If you come across an area where the threshold decreases, disappears completely, or increases slightly over a broad area, it should be investigated. Since the changes will often be barely noticeable, it is worth practicing this technique before venturing into the field.

If you have access to some black sand, place a cup or two in a plastic sandwich bag and use it as a practice sample. By varying the distance between the sample and the search coil and listening to the change in the audio threshold, you will quickly learn how to identify pockets of mineralized material as well as determine how concentrated they are.

By using this method you will be able to pinpoint several good locations that warrant further investigation in a relatively short period of time. As you remove the dirt and process it with a pan,

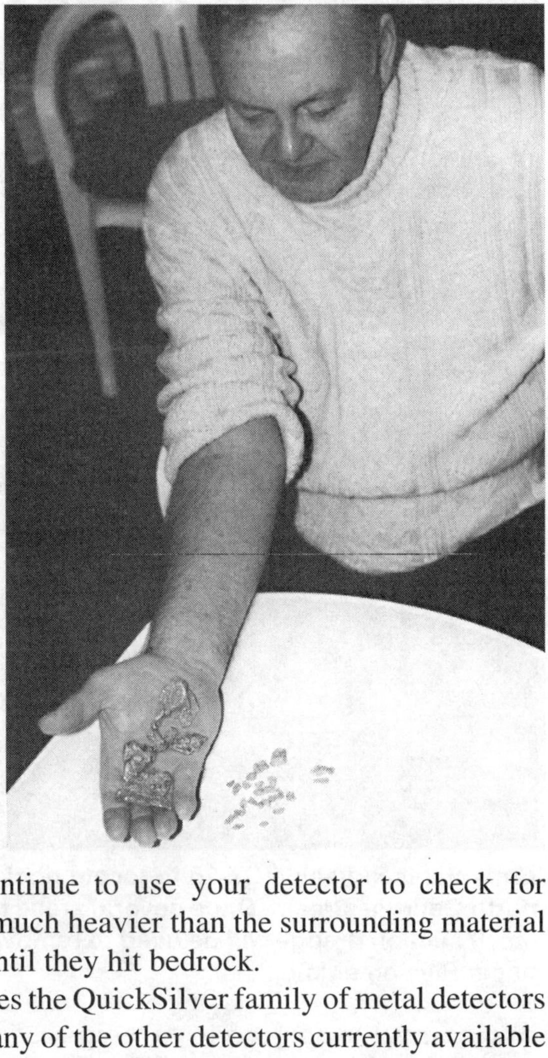

Jerry Denton displays almost 10 ounces of coarse gold nuggets recovered from a dry riverbed in the Yukon Territory by long-time mining associate Gerald Ahnert using a CZ-20.

sluice or dredge, continue to use your detector to check for nuggets, as they are much heavier than the surrounding material and will often sink until they hit bedrock.

A feature that gives the QuickSilver family of metal detectors an advantage over many of the other detectors currently available for electronic prospecting is their accurate discrimination circuit. One use was discussed earlier; however, it also allows treasure hunters to purchase one detector for both prospecting as well as coin and relic hunting rather than having to buy a different detector for each purpose. A number of seasoned prospectors have told me they often find old cabins and mine shafts during their

Andy Sabisch uses a CZ-20 to locate pockets of black sand in a north Georgia stream. Once several areas have been identified, a small suction dredge will be used to remove the sand, gravel and hopefully some gold.

searches, yet are forced to give up searching them due to the amount of trash normally found in these areas by their non-discriminating detectors. By packing in a CZ, one can switch over from prospecting to relic hunting by simply turning the discrimination knob to the appropriate position. This versatility will help ensure that the times you come home empty-handed become less and less frequent.

CHAPTER 7
COMPETITION HUNTING

Competition treasure hunts have been around for more than 25 years, and while the attendance at these events is usually quite good, there are a great many treasure hunters who have never participated in or even watched one. Organized treasure hunts allow hunters from different areas to get together and share a common interest while experiencing the excitement of locating coins and other valuable targets. It also provides treasure hunters with the opportunity to see other metal detectors and accessories on the market and pick up tips on how to use their own equipment more effectively.

In order to provide you with some of the tips and techniques that will allow you to improve your chance for success at a competition hunt, a discussion of how most hunts are conducted will prove useful.

The majority of organized hunts are one day in length, although there are a few weekend hunts that take place over a two-day period. Attendance can range from 20 or 30 entrants in a local hunt to 200-plus hunters in a large national hunt. The activities normally start early in the morning with registration, and the participants are encouraged to look over displays set up by local dealers and equipment manufacturers.

The first hunt is usually a "tune-up" hunt that allows everyone to get into the spirit of the day's activities. To begin the hunts, the participants line up around a large designated area and wait for a signal to start. Individuals responsible for organizing the hunt will have previously buried prize tokens and coins, including clad and foreign coins, wheat cents, Buffalo nickels, Indian Head pennies, silver dimes, quarters, halves and even dollars for the entrants to find. When the signal is given, everyone tries to recover as many targets as possible within the allotted time. Most hunts are 45 to 60 minutes in length with a 30-minute break in between.

The hunt activity will start to wind down by mid-afternoon.

At this time the trophies and prizes—which often include metal detectors, gold and silver coins, headphones, carrying bags and other accessories—will be awarded.

If you are interested in attending some of these hunts or if you already attend hunts on a regular basis, there are a few "tricks of the trade" that will help increase your overall success and enjoyment of the event. The first and foremost thing to do before going to a hunt is to be sure that your detector is in top operating condition. Replace the batteries before you start and then check their condition throughout the hunt. Bring along a spare set in case they fade on you during one of the hunts.

Headphones are a necessity at competition hunts simply because it's difficult to hear the signals from your detector when 100 detectors all around you are sounding off at the same time. A high-quality set of headphones (such as the Fisher Phones) have been designed to take the punishment treasure hunting subjects equipment to and will be worth the difference in price. Check your headphones before the hunt, paying particular attention to the wires in the plug, as this is the area where most problems originate. There is nothing more disappointing than to be forced out of a hunt due to a mechanical malfunction that could have been prevented by checking your equipment before leaving home.

If you have a backup metal detector, be sure you take it with you, so that you will not be out of the competition if your primary unit breaks down. This does not need to be a top-of-the-line unit. A quality detector such as a 1235-X, 1225-X or even the inexpensive 1212-X will be more than adequate and allow you to continue hunting in the event you have a problem with your primary detector.

If you have never been to a competition hunt, it would pay to set up a practice field to use before entering one. Lay out an area and bury a number of coins at depths ranging from 1 to 3 inches. Most of the targets in the actual hunt will not be buried much deeper than that since extensive damage to the ground would result if they were. Keep this in mind when participating in an actual hunt: if the organizers did not say that some coins would be deeply buried, DO NOT continue digging if the target is more than a few inches deep. Often I have seen hunters spend 5 or 10 minutes in a single hole and then wind up recovering a rusted horseshoe or some other large piece of trash. Remember, with thousands of coins to bury, the organizers do not have the time to bury them at depths of more than a few inches. Time is of the essence, so the less time you spend digging unproductive holes, the more coins and prizes you will be able to take home at the end of the day.

A good way to bury the coins in your practice area (which is also how most of the targets are planted in actual hunts) is to insert your knife into the ground at a 45-degree angle, remove the blade, drop a coin into the slit and step down on it to hide the hole. Once you have buried 30 to 50 coins, begin practicing by locating, pinpointing and recovering them. In order to be successful in a hunt, you must not only be able to locate targets quickly, you must also be able to recover them quickly. These practice sessions will allow you to develop a technique that works well for you while ensuring that you will not become tired halfway through the day's activities.

As you practice, you will find that you are able to recover more and more of the targets in the limited amount of time that you will have in the actual hunt. While this method of searching is not recommended when hunting for old coins in the local park, most of the targets you will be looking for are shallow, making speed and accuracy of pinpointing more important to your overall success. As your speed and ability improves, increase the size of the practice area so that you search it effectively in approximately 30 minutes.

When you are lined up and awaiting the start of the hunt, don't look at the hunt field as if you need to search the entire area in the time allotted. Hunt personnel will have buried an equal number of coins and tokens throughout the field so one area will not be any better than another. When the start signal is given, select an area equal to your practice area and hunt it thoroughly. After you have finished searching this area, you should simply wander throughout the hunt field attempting to locate targets that other hunters may have missed. Approximately 80 percent of all the good targets will be found within the first 30 minutes of a hunt, so random searching will probably produce as much as careful searching would in the final 15 to 30 minutes of the hunt.

One aspect of competition hunts that even some experienced treasure hunters forget about is how to properly set the various controls on their detector. With the large number of hunters in a small area, many detectors will experience electrical interference from each other. The easiest way to reduce or eliminate this problem is to lower your sensitivity setting. While this will reduce the depth at which you can detect a target, most of the targets will be quite shallow and a CZ can easily detect these targets even with the sensitivity control set at its minimum setting.

The discrimination control is also misused by most treasure hunters who enter a competition hunt. Many hunters use almost no discrimination to make sure they will not miss any targets. Because targets being buried are coins and tokens, sufficient discrimination (to eliminate iron, tinfoil and other common pieces of trash) should be used. Most hunts are held at sites that are used on a daily basis and usually contain a great deal of trash. Since you only have a limited amount of time, don't spend it digging worthless targets.

This is where the QuickSilver models truly excel over the competition. By setting the discrimination control at "4," virtually all trash targets will be rejected and only coins will produce an audio response. Due to the accuracy of the target-identification circuitry, you will still be able to detect nickels (Buffalo and V nickels are often buried in competition hunts) while ignoring pull tabs and screw caps found at most hunt sites. As a result, while other entrants are digging both good and trash targets, CZ users will be able to recover more coins and tokens in the same amount of time.

When hunting, make sure that your search coil stays close to the ground throughout the entire sweep. All too often, treasure hunters try to cover a larger area with each sweep only to find that the search coil is coming up off the ground several inches in an arc at the end of each sweep, resulting in missed targets. If you want to cover more area with each sweep, try replacing the standard lower rod that came with the detector with a longer one. The rod that comes with the CZs is 23 inches long; however, Fisher offers two lengths of replacement rods that can be used for this purpose. Available from your local dealer or directly from the factory, part

No. 202089-001 is 30 inches long and part No. 202089-002 is 37 inches. It is recommended that you try out the longer stem before attending as hunt, as the additional length will take a little getting used to.

Another important point to remember, particularly if this will be one of your first hunts, is to concentrate on the hunt itself and don't get distracted watching what other hunters are finding. During the hunt, the only person you are competing with is yourself, so just relax and focus on what your detector is telling you. At the first few hunts I attended years ago, I spent most of the time watching the other hunters rushing around the field recovering targets, and as a result, found very little myself. Remember, even if you find only one token, it could be the one that opens the treasure chest or wins a new metal detector.

The CZ-20 makes an ideal competition-hunt detector since hunts are usually held rain or shine. I have been to several hunts where it rained all day and many of the entrant's detectors broke down due to moisture by the end of the hunt. As a result, more targets remained for those treasure hunters who were able to continue hunting throughout the day. Many of the serious competition hunters I know around the country always take a waterproof unit along to the hunts they attend, usually as their primary unit. By hipmounting the control housing of the CZ-20, the overall weight of the armrest and search coil is reduced to less than 2.5 pounds, which prevents fatigue even after several hours of fast-paced hunting.

In summary, if there is an organized treasure hunt being held near where you live, consider attending it. The excitement you experience and the knowledge gained from talking with other participants is worth far more than the modest entry fee, and who knows, you might just wind up being the big winner for the event. One word of "caution" however: many treasure hunters have become addicted to competition hunts and often plan family vacations around hunts held in different parts of the country.

CHAPTER 8
ACCESSORIES

While any one of the QuickSilver metal detectors will produce outstanding results in the field, there are a number of accessories available that will improve your success in various situations as well as help keep your detector in perfect working order for years to come. This chapter describes some of these accessories and the benefit each provides.

Optional Search Coils

Currently, Fisher produces two optional search coils for the CZ-5, CZ-6 and CZ-6a models in addition to the standard 8-inch coil. As discussed in other chapters throughout the book, they are 5 and 10 1/2 inches in diameter. The CZ-20 can be configured with any of the three coils; however, since they are hard-wired to the control housing, users need to purchase the detector with the coil that will best fit their own needs.

As discussed in Chapter 1, CZ-5 and CZ-6a search coils are interchangeable; however, CZ-6 coils can only be used on CZ-6s. While a CZ equipped with the standard 8-inch coil will provide above-average performance in most cases, the use of an optional search coil can help locate targets that would have otherwise gone undetected.

In the past, most smaller coils resulted in a significant loss of detection depth, and as a result, few treasure hunters felt they were worth using. Fisher's 5-inch coil was specifically designed for the QuickSilvers to minimize any loss of detection depth while providing all of the benefits normally associated with a smaller coil. The two biggest benefits realized from using the 5-inch coil are reduced falsing or chattering from extremely mineralized ground and improved ability to detect good targets that are buried adjacent to trash targets. Several treasure hunters from around the country who use CZs exclusively in their searches provided the following examples of how the 5-inch coil put more finds in their collection.

- A well-known and heavily hunted Civil War campsite in Virginia produced seven Confederate buttons, a handful of bullets and three coins, including an 1854 half dollar in two days from an area filled with nails and other pieces of iron.
- A husband-and-wife team in New York City found 18 silver coins, 24 Indian Head pennies and a number of other early-date coins in a portion of Prospect Park that others had given up on due to the high amount of trash.
- A treasure hunter that specializes in searching Arizona ghost towns found several trade tokens, a brass skeleton key and a gold earring around two foundations that were heavily littered with trash.

The 10.5-inch coil offers advantages over the 8-inch coil in certain situations as well. First, one can cover more ground with each sweep (the 10.5-inch coil will cover about 75 percent more area than the 8-inch coil). In areas where targets are few and far between (such as a well hunted park, battle site or large ocean beach), the 10.5-inch coil will produce much better than the 8-inch coil. The larger coil also detects deeper targets. The only drawback to the larger coil is its additional weight; however, by sliding the control housing off the shaft and hunting in a hipmount configuration, the weight can be minimized.

So if you want to maximize your CZ's potential, add both the 5- and 10.5-inch coils to your equipment "arsenal"—you might just be surprised at what everyone (including yourself) has passed up in previous searches of "hard-hunted" sites.

Coil Covers

The primary purpose of the search-coil covers is to keep the bottom of the actual search coil from getting scratched or chipped when hunting over rocky or sandy terrain. Considering that a replacement search coil costs in excess of $100, purchasing a coil cover for $4 to $10 is an inexpensive investment that no one should be without.

There are two types of coil covers available for the CZ's search coils—an open and closed design. The open design looks like the spider coil itself and allows the user to see the ground beneath the coil, which aids in pinpointing targets. The closed design is more suited to relic hunters or those treasure hunters that routinely search overgrown locations. In these situations, an open coil cover can get snagged on branches and rocks, making it difficult to effectively search the area. The closed cover has been designed to eliminate this problem.

A word of caution for treasure hunters who use their CZs on saltwater beaches. Salt and black sand can build-up inside of the coil cover and adversely affect the detector's operation. It is advisable to remove the coil cover and rinse out any sand after each use under these conditions.

Headphones

The use of a quality set of headphones will probably affect your overall success more than any other accessory you can buy. Headphones will actually provide you with several benefits.

First, you won't bother others if you're "beeping" in a public area. Secondly, unless you are searching a site with no background noise (such as car traffic, wind, waves, children playing), many of the marginal signals that often indicate older, more-valuable targets might be missed due to the extraneous noise. Headphones will let you concentrate on what the detector is telling you, thereby allowing even the faintest signals to be detected.

Fisher offers three types of headphones designed for use on the CZ-5, CZ-6 and CZ-6a. The first is a basic set of headphones featuring individual volume controls for each ear and padded ear cushions. Simply called Stereo Headphones (part No. 972095) they have been designed to withstand the punishment of continuous outdoor use.

The second type of headphones available are called Fisher Phones, and their design blocks out virtually all outside noise. Featuring replaceable ear pads; a built-in circuit that limits the strength of the audio signal produced by large, shallow targets; individual volume controls for each ear; moisture resistant construction; and a 5-year warranty, Fisher Phones are a wise investment if you are serious about finding treasure. Despite the name, Fisher Phones can be used on other brands of detectors with the same performance enhancements being noted.

The third type of headphones that were designed strictly for the CZ-6 and CZ-6a user are called simply Waterproof Headphones. As the name implies, these headphones are waterproof and ideally suited for treasure hunters who search the shallow-water areas of swimming beaches or relic hunters that often find themselves miles from their vehicle when the rain starts. They are ruggedly built and designed to provide years of dependable service.

Carrying Cases

Carrying cases not only provide a convenient means for carrying your detector and related accessories, they help protect them from unnecessary wear-and-tear.

Fisher offers two different styles of carrying cases, a soft vinyl bag and a hard-sided case that resembles a small suitcase. The vinyl case, designed to fit all of the QuickSilver models, allows the detector to be placed in the case without requiring any disassembly (i.e., removal of the search coil or separation of the shaft). There is enough room inside to store a set of headphones, small digging tool, a pouch to hold your "goodies," and even an extra search coil in addition to the detector. It also features a shoulder carrying strap which is useful if you are planning on packing the detector into a remote site. While offering limited protection from unexpected bumps, it will keep dust and moisture away from the detector.

If additional protection is required, the hard-sided carrying case is a recommended option. Featuring locking clasps and a foam-filled interior with cut-outs designed to fit the detector, it will protect your detector from virtually any bumps or jarring that might be experienced while traveling. While some disassembly is required to fit the detector in the case, the protection it affords is well worth the few minutes it takes to remove and re-install the search coil.

The CZ-20 comes standard with its own hard-sided, lockable carrying case.

Miscellaneous Items

- **Cordura® Control-Housing Holster:** Designed to be used when hipmounting the control housing of the CZ-5, 6 and 6a, the holster helps protect the housing from dirt, dust and scratches. Reasonably priced, the holster is a worthwhile accessory if you spend much time hunting with your detector in the hipmount configuration.

- **External Speaker:** For use only with the CZ-6 and 6a, this accessory plugs into the headphone jack and mounts on the top of the control housing. Since the CZ-6 does not have a built-in speaker, it is designed to allow users to hear audio signals without the need for a set of headphones. Considering it is not rain or splash-proof and limits the user's ability to hear faint signals as with most built-in speakers, this is an accessory more suited for demonstrating how the CZ-6 responds to targets rather than for use in serious treasure hunting applications.
- **Spare Lower Shaft:** As described in the chapter on competition hunting, Fisher currently offers replacement lower shafts in three lengths: 23, 30 and 37 inches. While the standard, 23-inch length is more than adequate for most users, there are times when the longer shafts can come in handy. While the most obvious application would be for a tall treasure hunter, coin and beach hunters can benefit from using a longer shaft as well. By replacing the standard-length shaft with one of the two longer ones, more area can be covered with each sweep. This is only recommended for open, grassy areas or the wide expanses of sand found on most beaches, as the increased length makes it difficult to work around trees, rocks and other obstructions. A number of seasoned treasure hunters have told me that they keep each of their optional search coils mounted on a lower shaft to reduce the amount of time it takes to switch coils in the field. By having the coil attached to the shaft and the wire secured with a pair of Velcro® strips, switching coils is simply a matter of removing one lower shaft and re-installing the one with the desired coil on it. Not only is this a real time saver, but it helps reduce the possibly of damaging the coil wire caused by bending and twisting it each time the coils are to be switched.
- **"Save-a-Hunt" Kit:** Nothing can be more frustrating than driving 50 miles to a promising site, starting to recover targets,

and then having your detector break down. What a number of successful treasure hunters do to prevent this from happening is bring along a "Save-a-Hunt" kit, which allows them to correct most minor problems that can put an end to a day's hunt. Kept in a small plastic box, the kit contains the following items:

- **Spare Batteries:** Since you never know how long you might decide to stay at a site that is producing well, having a spare set of batteries with you will save the time needed to find a store that sells batteries in the event yours die while in the field.
- **Battery Holder:** While rare, the battery compartment on the CZ-20 can flood, resulting a damage to the battery holder. Having a replacement with you will save a day that would otherwise have been lost. The part number for the CZ-20 battery holder is 2037582.
- **Search Coil Washers:** Frequently the rubber washers that help hold the search coil in the desired position become worn and no longer perform their designed function. This is especially true on detectors used in the water or on the beach, as sand acts as an abrasive and can quickly ruin a set of washers.
- **Tools:** By including a pair of needle-nose pliers, a few screwdrivers and a roll of electrical tape in the kit, you will be able to make repairs in the field with a minimum amount of down-time.

* CHAPTER 9
THE NEW QUICKSILVERS: CZ-7, CZ-7a and CZ-7a Pro

In the short time that has elapsed since this book was first published, Fisher has moved ahead rapidly with its very successful QuickSilver line of metal detectors. The digital CZ-7 replaced the CZ-6a and then the CZ-7 was replaced by the CZ-7a. A body-mount version of the CZ-7a has been added, the CZ-7a Pro. The current lineup consists of the CZ-5, CZ-7a, CZ-7a Pro and CZ-20.

The Fisher CZ-7 QuickSilver was the first Fisher detector to abandon most analog knobs and meters in favor of microprocessor-driven touch-pad controls. The CZ-7 control panel features a liquid crystal display with a cursor to select various search settings.

* *Chapter 9 was written by David Toussaint, Marketing Publications Writer, Fisher Research Laboratory.*

CZ-7

The most significant change to the QuickSilver line came with the introduction of the CZ-7, which replaced the CZ-6a. Instead of an analog target-I.D. meter complete with meter needle and icons (see p. 5), the CZ-7 featured a totally reworked control panel that uses microprocessor touch-pad controls and a liquid crystal dis-

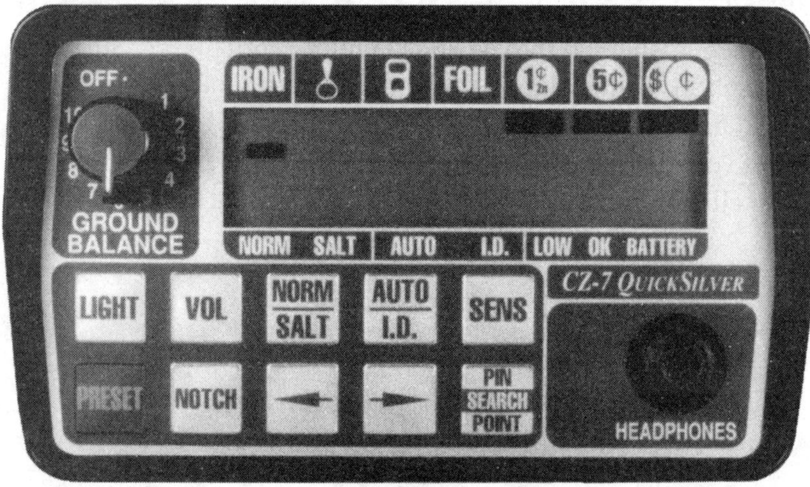

The anolog meter and control knobs of the CZ-6 (see p. 5) were replaced by a cursor, touch-pad controls and a liquid crystal display on the CZ-7.

play to identify targets, monitor battery condition and display control settings.

The touch-pad, microprocessor controls of the CZ-7 also allow for a new, notch discrimination system. Unlike the CZ-20, CZ-5, CZ-6 and CZ-6a, the CZ-7 offers true "notch" discrimination, meaning that any individual target category (or group of categories) can be selectively accepted or rejected. For example, if the operator wants to hunt in a search mode that rejects pull tabs and accepts iron, that's fine with the CZ-7. The operator uses the arrow touch pads to move the cursor under a target icon and the NOTCH touch pad to accept or reject it.

In earlier CZ models, the operator could only select a target category for rejection if the targets below it were also rejected. For example, the CZ-5 won't reject pull tabs without first rejecting iron, and foil can't be rejected without first rejecting pull tabs. However, with the microprocessor controls of the CZ-7, the operator can reject any target category or group of categories without worrying about where it falls within the discrimination hierarchy.

This feature is particularly useful for serious gold-ring hunters who want to dig only those targets that are most likely to be gold rings. Knowing that most gold rings fall into the tab and foil categories (see p. 49), these folks will notch out coins and iron so their CZ will respond only to the mid-tone foil and tab targets. Coin hunters may choose to notch out zinc pennies so they can spend more time digging older, more valuable coins. Relic hunters may determine that a particular target category, or group of categories, is just what they need to find a particular type of relic.

The one analog knob that remains on the CZ-7 is Ground Adjust. All other adjustments—like sensitivity, volume, notch and mode—are controlled by the touch pads and LCD. The touch-pad controls are tactile-feel, meaning that they click slightly when depressed, letting the operator know instantly if the touch-pad

The second generation CZ-7 QuickSilver, the CZ-7a, includes all the features of the CZ-7 in addtion to a weather-proof speaker, reduced battery drain, faster pinpointing and new software.

switch was activated. Depth reading is activated by a touch pad that displays depth in inches on the LCD.

Another first for the CZ-7 is on the inside. Sate-of-the-art surface-mount electronics offer size, weight and reliability advantages over traditional solder-board assembly methods.

CZ-7a

The CZ-7a, released after the CZ-7, improved on features of the CZ-7 and added some features of its own. Like the CZ-7, the CZ-7a automatically returns to the last setting before shutoff, but some slight battery drain during shutoff (a characteristic of the CZ-7) was eliminated in the CZ-7a. At turn on, the CZ-7a returns to the last setting used before shutoff, but battery drain during shutoff is zero.

Most notably, the CZ-7a saw the return of the built-in speaker, this time a sealed, all-weather speaker, making the detector weather proof. As with the CZ-7 , the CZ-7a is not hip mountable.

Also reintroduced by the CZ-7a was the round, double locking stem. Some earlier models used a stem with a flat side to keep the search coil straight. But Fisher decided the original, round stem was the way to go with the CZ-7a.

CZ-7a Pro

At first glance, the CZ-7a Pro doesn't look much different than the CZ-7a, but on closer inspection, you'll see that the battery compartment is missing and the control housing is mounted on a pistol-type grip. The changeover to surface-mount production allowed enough room in the newly designed control box to house two, 9-volt batteries.

Consequently, the CZ-7a Pro is lighter (3.8 lbs.) than the CZ-7, and has a lot less battery life (15-20 hrs.).

Most significantly, however, the CZ-7a Pro can be body mounted, an important feature for anyone wanting to lighten the load on their arm or hunt in shallow water. The control housing slips off the pistol grip and attaches easily to a special belt clip, which comes with every CZ-7a Pro. Once the housing is on your hip, what's left to swing (the handle and coil) weighs only about

The CZ-7a Pro is the latest entry by Fisher Research Laboratory into the treasure hunting market. It's lighter than its predecessors and features a hipmountable control housing.

2.5 lbs.

Like the CZ-7, the CZ-7a is rain proof. To make the detector rain proof when the headphones aren't plugged in, a small plastic plug was attached to the input jack of the Pro model. This plastic piece can be inserted into the headphone jack to make the CZ-7a Pro completely rain proof when the headphones aren't in use.

The CZ-7a Pro represents the latest, greatest offering from Fisher Research Laboratory. Like all the previous QuickSilver models, the CZ-7a Pro is noted for its depth penetration, target-I.D. accuracy and ease of use.

Summary

Note: Fisher now offers a chest harness especially for Fisher body-mount detectors. It's a great accessory for shallow-water hunters.

Fisher practices a policy of continuous product improvement. Problems are corrected, customer suggestions are implemented, new technology is utilized—constantly. Most changes are minor and invisible to the user. More important upgrades that do not affect the basic operation of a detector may result in the addition of an "a" to the model name. (The CZ-6a replaced the CZ-6 and the CZ-7a replaced the CZ-7.) A major change, such as the addition of "notch" circuitry, may result in the birth of an entirely new model. (The CZ-7 replaced the CZ-6a.) And a variation, such as a new body-mount feature, might result in the addition of something like "Pro" to the model name. (Note that the CZ-7a Pro did not replace the CZ-7a; it was an addition to the line that gives the user a choice between very long battery life and the ability to body-mount the same detector.)

The table below illustrates some of the basic differences between Fisher's current CZ-QuickSilver models.

FEATURE	CZ-5	CZ-7a	CZ-7a Pro	CZ-20
1. Target I.D.	3-Tone + Meter	3-Tone + LCD	3-Tone + LCD	3-Tone
2. Backlight	No	Yes	Yes	—
3. Rain proof	No	Yes	Yes	Yes
4. Submersible	No	No	No	Yes
5. Notch	No	Yes	Yes	No
6. Battery hours ①	15-20	80-90	15-20	35-55
7. Touch pads	No	Yes	Yes	No
8. Body Mount	Yes	No	Yes	Yes
9. Speaker	Yes	Yes ②	Yes ②	No ③

① Alkaline batteries ② Rain proof speaker ③ Water proof headset

RECOMMENDED READING LIST

Treasure Hunting With a Metal Detector In, Around, and Under Water by Steve Blount and Lisa Walker

Advanced Shallow Water Treasure Hunting with the Fisher 1280-X Aquanaut by Wallace Chandler

Advanced Water Working Techniques by Andy Sabisch

Advanced Nugget Hunting with the Gold Bug Metal Detector by Pieter Heydelaar and David Johnson

How to Find Gold Nuggets in Dredge Tailings by Gerald T. Ahnert

The Poor Man's Treasure Hunter by Bob Trevilian and Frank Carter

Treasure Laws of the United States by Roger Grimm

The Detectorist by Robert Sickler

The Complete Book of Competition Treasure Hunting by Ernie "Carolina" Curlee

A New Guide to Treasure Hunting by H. Glenn Carson

America's Lost Treasures by Micheal Paul Henson

Civil War Collector's Price Guide by the editors of North & South Trader's CIVIL WAR Magazine by North & South Trader

Search! by Jim Warnke

Coinshooting Vol. I, II and III by H. Glenn Carson

Underwater Treasure Hunting Made Easy by Carl Fismer (a 90-minute videotape providing an introduction to underwater treasure hunting)

TREASURE HUNTERS' CODE OF ETHICS

- I WILL always check federal, state and local laws before searching. It is my responsibility to know what the applicable laws are.
- I WILL respect private property and do no treasure hunting without the owner's permission.
- I WILL fill all excavations.
- I WILL appreciate and protect our heritage and natural resources, wildlife and private property.
- I WILL use thoughtfulness, consideration and courtesy at all times.
- I WILL build fires in designated or safe places only.
- I WILL leave gates as found.
- I WILL remove and properly dispose of any trash that I find. I will not litter.
- I WILL not damage, deface, destroy or vandalize any property, buildings or what is left of structures.
- I WILL not tamper with signs, structural facilities or equipment.
- I WILL assist law enforcement agencies when possible.
- I WILL return valuables to their rightful owner whenever possible.
- I WILL report the discovery of any items of historical significance to the local historical society or proper authorities.
- I WILL promote responsible historical research and artifact recovery, and the sharing of this knowledge with others.

Federation of Metal Detector and Archeological Clubs

The Federation of Metal Detector and Archeological Clubs is a national organization whose primary goal is to preserve and protect the rights of metal detectorists. To obtain more information about this organization, or to find out how to join, contact the FMDAC at the following address:

Federation of Metal Detector and Archeological Clubs
431-80 Ave. N., St. Petersburg, FL 33702

About the Author

Andy Sabisch began his treasure hunting career over 30 years ago when his parents gave him an inexpensive metal detector as a birthday gift that was ordered from the back of a magazine. They figured that it would go the way most toys do for kids that age—relegated to the back of a closet after a few weeks. Surprisingly, Andy and his brother Chris stuck with it and hunted schools, parks and private yards surrounding their home just outside of New York City. Over the next year their efforts yielded a total of 26 coins and an untold amount of trash. Unwilling to give up, he upgraded his equipment and began finding an increasing number of coins and other valuables from sites that had never been searched before.

Andy's interest in treasure hunting has remained high throughout the years. Travel, both personal and business-related, has allowed him to try his hand at many different types of hunting around the world. These include wreck diving along the East Coast and Caribbean; prospecting for gold in the Northwest states and Alaska; detecting many of the old ghost towns in Idaho, Montana and Wyoming; hunting for bottles in New England; searching for artifacts from both the Civil and Revolutionary Wars; coin hunting throughout the United States, Mexico, Europe and South America; and attending well over 150 competition hunts.

Andy is always willing to share the knowledge he has gained over the years with fellow treasure hunters. Since his first article was published in 1982, he has had nearly 600 articles published in various treasure hunting and diving magazines. In addition to the articles, he has written several books that have received widespread acclaim from people around the world and has hosted dozens of seminars introducing treasure hunting to groups ranging from the Boy Scouts to amateur radio clubs and senior citizens.

It's funny in a way that something that started out as what was expected to be nothing more than a passing fancy has remained a central part of Andy's life. Since his wife, Rosanne, and 5-year-old son, Paul, also share in his interest, many of their family trips are planned around some form of treasure hunting.